The Art and Science of Poisons

Authored by

Olen R. Brown

Dalton Cardiovascular Research Center, University of Missouri,
Columbia, MO, USA

The Art and Science of Poisons

Author: Olen R. Brown

ISBN (Online): 978-1-68108-697-2

ISBN (Print): 978-1-68108-698-9

© 2018, Bentham eBooks imprint.

Published by Bentham Science Publishers – Sharjah, UAE. All Rights Reserved.

First published in 2018.

General:

1. Any dispute or claim arising out of or in connection with this License Agreement or the Work (including non-contractual disputes or claims) will be governed by and construed in accordance with the laws of the U.A.E. as applied in the Emirate of Dubai. Each party agrees that the courts of the Emirate of Dubai shall have exclusive jurisdiction to settle any dispute or claim arising out of or in connection with this License Agreement or the Work (including non-contractual disputes or claims).

2. Your rights under this License Agreement will automatically terminate without notice and without the need for a court order if at any point you breach any terms of this License Agreement. In no event will any delay or failure by Bentham Science Publishers in enforcing your compliance with this License Agreement constitute a waiver of any of its rights.

3. You acknowledge that you have read this License Agreement, and agree to be bound by its terms and conditions. To the extent that any other terms and conditions presented on any website of Bentham Science Publishers conflict with, or are inconsistent with, the terms and conditions set out in this License Agreement, you acknowledge that the terms and conditions set out in this License Agreement shall prevail.

Bentham Science Publishers Ltd.
Executive Suite Y - 2
PO Box 7917, Saif Zone
Sharjah, U.A.E.
Email: subscriptions@benthamscience.org

BENTHAM SCIENCE

CONTENTS

FOREWORD

The author has written a superb review on the toxicology of numerous poisons that in general are not readily obvious. These include for example, arsenic, hemlock, scorpion venoms, plants, poisons that are associated with sea habitants, spiders, and snakes. Each chapter reviews in detail the differences in poisoning by these species. Finally, for those with a scientific background the author provides an excellent review of a potpourri of agents which can be toxic depending on the dose ingested or administered, and as the author points out in chapter 1, in the words of Paracelsus, "the dose makes the poison". In the chapter on a potpourri of poisons the author discusses these chemical agents in detail including their mechanism of action. These include several drugs used clinically, such as opioids for pain, statins for treating hypercholesterolemia, doxorubicin for the treatment of cancer, curare, a muscle relaxant but which has been replaced by newer agents, warfarin an anticoagulant, the various alcohols, and carbon monoxide which is responsible for many emergency room visits. Finally, oxygen toxicity which may surprise many readers is discussed in detail as this has been the author's area of research interest for decades. This last chapter entitled "a potpourri of poisons" will most likely require a background in chemistry and biology. This should not deter non-scientists from reading this book. In case of anything, it may convince non-scientists to consider a career change. The greatest strength of this book is that the author has provided one source a detailed compendium of appropriate internet references which allows the reader to obtain further knowledge on that specific poison. The book targets an audience that is generally interested in toxicology but not necessarily requiring a detailed background in the basic sciences, although some exposure to chemistry and biology would be helpful. This is a good book to have in home, particularly regarding the discussion of poisonous plants, aquatic species, spiders, and snakes. This should also be a good reference source for those working in state and federal national parks. The book is well-written and easy to read by a non-scientist, except for the last chapter, a potpourri of poisons which does require knowledge in chemistry and biochemistry. The book has a wealth of useful information. The book should also serve as a useful text for undergraduate toxicology programs. The author's inclusion in several chapters of his own personal exposure to some of these potential poisons during his adolescent years provides a feeling to the reader of being there. The figures presented compliment the text and make the reading interesting and the readers desire to read more. The real strength of this book, however, is the remarkable extensive referencing provided by the author. He has produced a text with all of these references in one place for easy access for further readings. Well done Dr. Olen Brown.

Morris D. Faiman,
Department of Pharmacology
School of Pharmacy,
Life Span Institute
University of Kansas,
Lawrence, Kansas,
USA

PREFACE

I hope to entice the scientist and other readers of this book in equal measure. Poisons have two stories to tell. The science of poisons deals with the chemistry of toxic agents and the way they work at the cellular and molecular level. The art of poisons encompasses everything else about these agents that congers up the image of the skull and crossbones.

The science of poisons takes us on a voyage into the sub-microscopic world of atoms, molecules, and cells. Only there can we see the true miracles and mysteries of life and death. The mere existence of poisonous substances and especially the uses made of them by plants and animals are wondrous. Poisons are also used to explore the biological mechanisms of the body, to lower cholesterol by blocking its synthesis, to kill cancer cells, to destroy pests of all kinds, and as weapons of war. Science is neither moral nor amoral, only its uses can be so characterized.

The art of poisons encompasses everything else about poisons. It is the legends and stories of intrigue and murder and other deeply deplorable uses of toxic agents often with a surreptitious and evil intent. Let us hope that the future extends the beneficial applications of poisons and quells their evil uses.

A traditional, central concept in toxicology can be stated simply: the lethal dose of a substance is the amount required to kill the average person (the LD50). Today for most poisons, the mechanism of how they kill is known at the molecular level. Therefore, I propose that a new measure of toxicity based on the number of molecules required to kill (the $LD_{50}*$) is appropriate.

The simplest summary idea about poisons is one of the oldest—the dose makes the poisons (paraphrased from Paracelcus). I believe the most extreme example is the toxicity of oxygen. Oxygen is essential; we cannot live more than a few minutes without oxygen; however, it is detectably toxic at approximately two times the concentration found in air, and at hyperbaric pressures it is lethal.

Olen R. Brown
Dalton Cardiovascular Research Center
University of Missouri
Columbia, MO
USA

CONSENT FOR PUBLICATION

Not applicable.

CONFLICT OF INTEREST

The authors declare no conflict of interest, financial or otherwise.

ACKNOWLEDEGEMENTS

This book was written out of my experiences gained through scientific study, experimentation, and life. I was aided by the many individuals in my laboratory and the students I was privileged to instruct at the University of Missouri. I also thank my students and professors at the University of Oklahoma where I learned a deeper appreciation for science. I thank them all with deep gratitude borne out of experiencing the grace brought by science and discovery.

I especially thank Claire Engler and John Allen for Art work illustrating concepts in the book and Cameron Brown for invaluable assistance with preparation of the manuscript and computer formatting.

DEDICATION

To Cecilia and Stella.

CHAPTER 1

The Deadliest Poison

Abstract: Beauty is said to be in the eyes of the beholder. Likewise, the deadliness of a poison depends on subjective criteria. Is more weight to be given for quickness of action, stealth, whether an antidote is available, or how little is required? Most accounts declare botulinum A, the toxin produced by a species of anaerobic bacteria, to win the contest based on its LD_{50} (the amount that kills half of those exposed). Its toxicity, measured this way, is greater than any known substance. I propose a new way of ranking poisons, the LD_{50*}, based on the number of molecules in a deadly dose. This is more equitable because poisons differ greatly in their molecular weights – some are very small and some are very large molecules, and poisons kill molecule-by-molecule. Several snake venoms are deadly and the most toxic is that of the inland taipan, although the coral snake and cobras have very toxic venoms, and rattlesnake, because of the volume injected and the multiplicity of toxic ingredients, deserve mention. Only two species of scorpions (the death stalker and man killer; neither found in the United States) have stings that are life threatening for humans. Spider venoms don't quite make it to our most deadly list. Radiation exposure is a different kind of "poisoning" and Polonium-210 makes our list because of the small amount required and its intense radiation based on its very short half-life. The most deadly, quick-acting toxins affect the nervous system and cessation of respiration or heart function stops the supply of oxygen to tissues to cause death. It is thought-provoking to consider that all things are poisonous, and that only the dose makes the difference (to paraphrase toxicologists). In this context, life-giving pure water becomes deadly when several liters are consumed rapidly. Why poisons exist has a scientific answer, but perhaps not a satisfactory philosophical answer.

"All things are poison and nothing is without poison; only the dose makes a thing not a poison."[1]

Keywords: Acetylcholinesterase, Anaphylaxis, Antidote, Antitoxin, Antivenin, Avogadro's Number, Bane, Biological Warfare, Botulinum A, *Clostridium Botulinum*, Coral Snake LD_{50}, LD_{50*}, Myoneural Junction, Nerve Gas, Paracelcus, Poison, Polonium-210, Potion, Sarin, Skull and Crossbones, Taipoxin, Toxin, Venom, VX Agent.

INTRODUCTION

The word poison, to me, conjures up thoughts of death, skull and crossbones, chemicals, harm by intrigue and stealth, serpents, object of aversion, and when thinking technically, an inhibitor of a chemical reaction, especially an enzyme. Although we may not usually consider medicines (including the drugs known as statins that are used to treat elevated cholesterol) as toxic, many actually poison enzymes in our bodies. Merriam-Webster [1] gives several meanings for poison including that it is "… a substance that through its chemical action usually kills, injures or impairs an organism". This dictionary states that the origin of the word is from Middle English (derived from Anglo-French meaning *poisun drink*, potion, poison); or from the Latin for *potio*. Synonyms are bane, toxic, toxin, and venom. Bane is seldom used now but usually means a cause of great distress or annoyance, but can refer to a deadly poison, and conveys other meanings: killer, slayer, death, destruction and probably dates back to the 12th century. These meanings will define the scope of this book: stories of discovery, intrigue, murder, venomous animals, toxic bacteria, beautiful but deadly plants, and warfare by mankind.

POISONS MOST VILE

Paracelsus (1493-1541), credited with originating the discipline known as Toxicology (the study of poisons), is famous for having penned what is most often paraphrased simply as "the dose makes the poison". This provides very wide latitude for the subject of this chapter and I will take into consideration much more than how little of a substance is required to kill a human. Thus, we shall approach the question of the most toxic substance from a variety of perspectives.

Candidates of poison which kill a human with the least amount of substance traditionally have included botulinum [2] the protein produced by the microbe that causes botulism [3], the radioactive element polonium-210 [4], the venom of poisonous snakes including the taipan, cobra, mamba, saw-scaled viper, and others [5] and the nerve agents which have been used as weapons of war [6]. Use of poisons is often considered insidious and secretive; some are deadly but slow to act and some do their deadly work quickly. Some poisons always, or most usually, are encountered accidentally and others mostly by evil design.

Poisons are connected with murder by stealth and poisoning has been considered a cowardly act. John Fletcher (1579- 1625) is quoted to have used the words: "The coward's weapon, poison" [7]. Fletcher was a dramatist and wrote and collaborated to write comedies and tragedies. Unfortunately he died in the London plague of 1625 and it has been said that this resulted because he lingered in the city to be measured for a suit of clothes instead of quickly fleeing to the country as did many [8].

The universal symbol for a poisonous substance is the skull and crossbones (Fig. **1**). The origin of this is lost in history. However, I found the following information (with original attributions listed) [9]. I quote from this site as follows:

> "… it was first seen on the tomb of Tutankhamen… It is widely believed that the skull and crossbones was first used by the Knights Templar… in the Middle Ages… According to Masonic legend, the skull and crossbones are the bones of Jackes de Molay… Toward the end of the Roman Empire and into the Middle Ages, Christians frequently used the Skull and Crossbones to symbolize death… Originally, pirates used a red flag on the top of their ships' mast… many pirates soon changed their flags from red to black, and began to weave the skull and crossbones into them… Yale University formed a secret society called "Skull and Crossbones"… The Death's Head or *Totenkopf* was a symbol of the Nazi *Schultzstaffel* (SS: literally defense echelon, especially Hitler's body guard)… In 1829, New York State Law was changed to require that all containers of poisonous substances be labeled… The skull and crossbones first illustrated those labels in 1850."

Fig. (1). Universal symbol of a toxic substance.

Ranking of Poisons by Lethal Dose

It has become customary to compare the lethal amounts of poisons based on weight (mass). By this measure it is generally agreed that botulinum toxin has the smallest lethal dose of all poisons [2, 10 - 12]. Taipoxin (from the inland taipan snake), is second and it is the most potent of all venoms [13]. Third is polonium-210, which is extremely unstable (radioactive), and it is often cited as the most dangerous radioactive element [4, 14, 15]. Fourth in toxicity by required amount is the chemically-synthesized nerve poison VX agent (developed for warfare). It is the most toxic man-made chemical with a lethal dose that is difficult to accurately assess. Ranking poisons is problematic and will be a struggle faced throughout this book. Based on the sources cited directly above, my judgment is that the toxicities (per kg of body weight), of these agents are: type A botulinum toxin: 1.0 nanogram, polonium-210: 11.1 nanograms, VX-agent: 143 micrograms, and taipoxin: 2 milligrams. Note the difference in units: nanograms, micrograms, and milligrams; each is a thousand times smaller than the next, respectively. Some comments are necessary. The toxicity for taipoxin (the best source is [13]) is in the range of 0.01 mg/kg to 0.025 mg/kg for the complete venom but 2 mg/kg for taipoxin which is the presynaptic neurotoxin (surely the most toxic component). The difference in toxicity is in the range of 80-fold to 200-fold! Thus, if the data are correctly reported, the answer to this quandary may lie in the complex interaction of the venom molecules. A sophisticated study of the venom by Laura Cendron and coworkers provided data supporting several relevant conclusions [16]. Venom of the Australian taipan snake contains taipoxin, a neurotoxin with phospholipase A_2 activity that is a potent inducer of paralysis through the specific disruption of the neuromuscular junction pre-synaptic membrane. They state: "Although no correlation has been reported between neurotoxicity and enzymatic activity, toxicity increases with structural complexity and phospholipase A_2 oligomers show 10-fold lower LD_{50} values compared to their monomeric counterparts." This strongly indicates a correlation between structural complexity of the toxin and its neurotoxicity. They further state that their research revealed "... there are two isoforms of the taipoxin β subunit which show no neurotoxic activity but enhance the activity of the other subunits in the complex". In the introduction to their paper, they state: "Taipoxin is composed of three homologous non-covalently bound class I secretory PLA_2 subunits... We report the structural characterization of two isoforms of the taipoxin subunit β, which retains no enzymatic activity but is crucial for the elevated neurotoxicity of taipoxin. Moreover, we propose a novel model for the quaternary structure of a trimeric PLA_2 neurotoxin that has been determined under physiological conditions."

Thus, it appears that this toxin is more than the sum of its parts. Indeed, study of toxins at the molecular level is aided by considering their toxicities on a molecular basis and the LD_{50}* that we propose appears to be a helpful approach.

Scientific assessment of toxicity for humans is complex for many reasons which I will address near the end of this chapter. For now, let us compare, for four poisons that I calculate are the most toxic, the lethal amount, time required to kill, and availability and effectiveness of an antidote (Table **1**). Note especially that I have calculated the lethal amount in two ways. The traditional way to assess toxic potency is according to dose by weight. I propose a comparison, based not on weight but on the number of molecules of the poison in a lethal dose.

I propose that it is better to compare the lethal dose in terms of the number of molecules (atoms for polonium-210) because poisons exert their effects molecule-by-molecule. Note that plonium-210 is an atom and not a molecule. The time to death and availability of an antidote also are attributes that I propose (Table **1**). They are arguably arbitrary but all are justifiably quite important.

For clarity of comparison, the LD_{50} of botulinum can be equated to one with the LD_{50} of the other agents calculated proportionately with the following results: botulinum = 1; polonium = 11.1; taipoxin = 2 x 10^6; and VX agent = 0.143 x10^6. Comparisons of LD_{50}* (based on number of molecules) are: botulinum = 1; taipoxin = 6.5 x 10^6; polonium = 8,000; and VX agent = 8 x 10^7. It is remarkable that botulinum retains its first place rank in both assessments in spite of the fact that it is by far the largest molecule among the toxins and therefore there are fewer molecules in a lethal dose (Table **1**). The size comparisons for these poisons are as follows (in Dalton units): botulinum = 150,000; Taipoxin = 45,600; VX agent = 267; and polonium-210 = 209. If we equate polonium-210 to 1 the comparison is easier to comprehend: VX agent is 1.28 times as large, Taipoxin is 218 times as large, and botulinum is 718 times as large as polonium-210.

When assessment is based on the time to death from a lethal dose, granting a great deal of uncertainty in reported values, the order is VX agent, taipoxin, botulism, and polonium (Table **1**).

Toxin from *Clostridium Botulinum*

Botulinum is the toxin in spoiled food that is produced anaerobically at alkaline pH by *Clostridium botulinum*, the microbe that causes botulism. There are seven known types (A-G) of the toxin which have slightly different molecular structures but which can be identified by laboratory tests. Botulism in humans is usually caused by types identified as A, B, E (and rarely by type F). The toxins are composed of amino acids linked together in a protein with a large molecular weight of approximately 150,000 Daltons for type-A toxin (Fig. **2**). This is huge, approximately 150,000 times the weight of one hydrogen atom [17]. Type-A toxin has 1,259 amino acids linked together by peptide bonds [18]. This large arrangement of atoms is poisonous by two chemical actions. One poisonous action results from a smaller part that is approximately 50,000 Daltons in size. A larger part (approximately 100,000 Daltons) of the molecule (Fig. **3**) is the component that actually binds to nerves and blocks impulse transmission to muscle fibers [19]. The estimated lethal dose for a human is 0.091 microgram. This is equivalent to only 365 billion molecules. This may appear to be a huge number; however, there are 602 billion trillion molecules in a quantity of the toxin equal to its molecular weight expressed in grams. A molecule of botulinum toxin has a weight of 150,000 Daltons (molecular weight expressed in grams). For comparison, the smallest known protein has only 20 amino acids, and weighs only 2,171 Da; it is found in the saliva of the Gila monster [20].

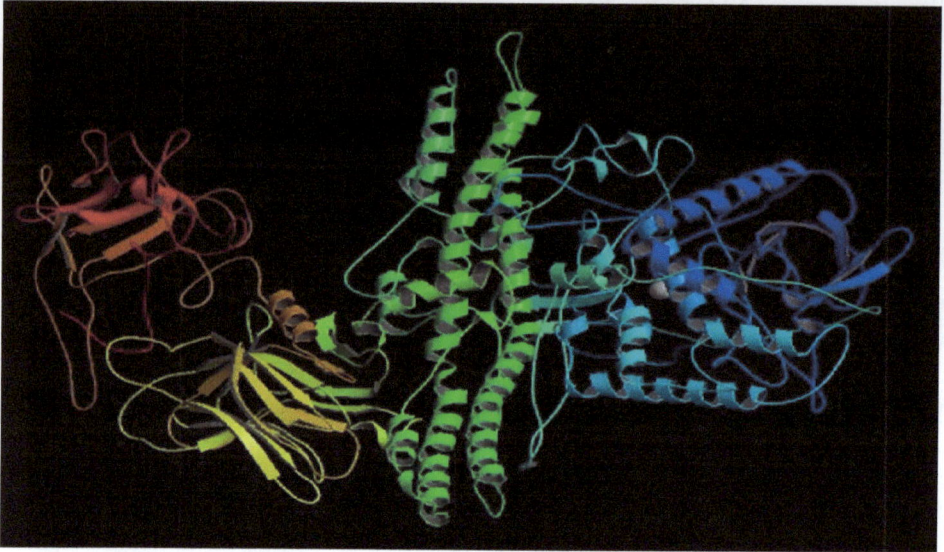

Fig. (2). Botulinum toxin, depicted as a "ribbon diagram" to show the complex 3-D structure of the protein.

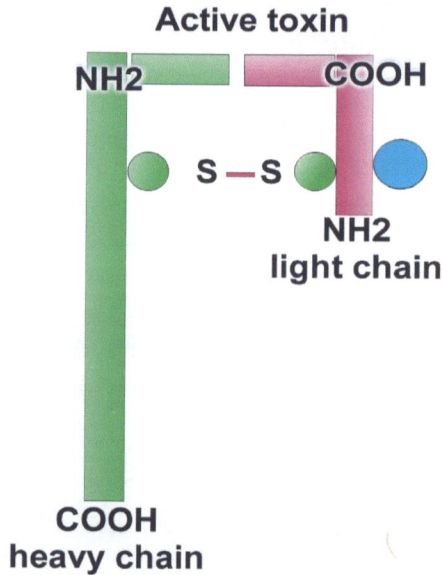

Fig. (3). Botulinum toxin, diagrammatically, showing two polypeptide chains linked by a disulfide bridge. The gap between the light and heavy chains (green and lavender) is the point of cleavage of the toxin at its site of cellular action into portions of 100,000 and 50,000 Daltons (see text).

The lethal quantity of botulinum toxin is a microscopic amount, far too small to be visible by the unaided, human eye. On a weight basis, nothing has been described to be more toxic for humans. This results directly from the fact that the toxin binds to nerves at junctions (synapses) and prevents nerve transmission. With a lethal dose, a person without treatment likely will die from paralysis of the breathing muscles. Paralysis develops first in muscles in the head area and descends. More specifically, botulinum toxin binds to receptors at myoneural junctions (connections between nerves and muscles) and prevents nerve impulse from activating muscle which causes flaccid paralysis. However, an antiserum is available as an antidote and if administered after rapid diagnosis, and with proper supportive care, the death rate is reduced to approximately 3% to 5% for clinical cases [10]. Breathing can be maintained by mechanical means if medical care is available and over time the body can replace the connections of nerves to muscles and patients can recover. However, this may require several months or even longer. A heptavalent botulism antitoxin (active against all seven known botulinum nerve toxin serotypes of *Clostridium botulinum*) has been developed but it is an effective antidote only when a proper diagnosis is made and treatment begun soon after ingesting the poison [21]. Damage that has already resulted from this toxin to nerves is not directly reversed by the antitoxin. After a lethal dose of botulinum death is unlikely before one day [10] and if the person seeks medical treatment, is properly diagnosed and given antitoxin (or mechanical ventilation and oxygen is available) the person likely will survive.

Nerve Gas

VX agent is the most deadly of the nerve gasses that have been synthesized in the laboratory for use as a terror weapon and in warfare (Fig. **4**). The estimated lethal dose by skin contact is 10,000 micrograms (about 1/100th of a milliliter by volume) [22]. Calculated in terms of the number of molecules, the estimated lethal dose of VX agent for a human (70 kg) by skin contact is approximately 22 billion billion molecules (compared to 280 billion for botulinum). Although the dose of VX agent is a large number of molecules, it is a very small amount. An ordinary "drop" is 1/20th of a milliliter; thus, divide an ordinary drop of liquid into 5 drops and one of these smaller drops is the amount of VX agent that kills by contact with the skin. Because the mechanism of poisoning is known and can be negated by an antidote which soldiers carry when there is threat of chemical warfare, and because sensors to detect VX agent are available, my assessment of the toxicity rank of this poison is decreased somewhat. This agent has a molecular weight of approximately 267 grams. This means, as is the case for all chemicals, the weight in grams is calculated based on the chemical composition of the substance. Small molecules are composed of fewer atoms and weigh less as a general characteristic. VX agent, therefore, is a comparatively small molecule. For comparison, water has a molecular weight in Daltons (molecular weight expressed in grams) of approximately 18; potassium cyanide is 65; aspirin 180, VX agent 267, table sugar 360, and botulinum 150,000.

The number of molecules in one gram molecular weight is approximately 6.2×10^{23} (Avogadro's number). This number is written as a power of 10 because it is so large. Written this way, the number of zeros is indicated by the exponent. It can be translated into more familiar terms as 6.2 hundred billion trillion. This is a number larger than the number of stars calculated to be in the entire known universe. Therefore, when we calculate the number of molecules present in the

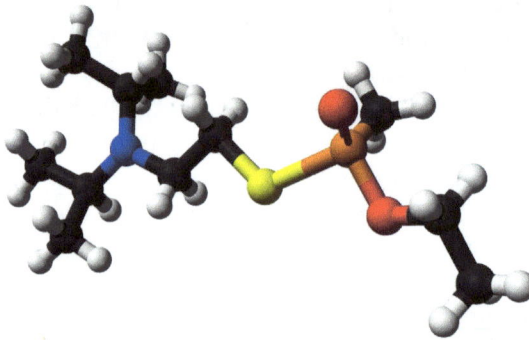

Fig. (4). VX agent shown as a ball and stick model. White is hydrogen, black is carbon, red is oxygen, orange is phosphorous, and blue is nitrogen.

lethal dose of a poison, even when a very small amount of the poison is lethal, the number is very large. Poisons vary greatly in molecular size. The action of poisons occurs at the molecular level, so calculations based on the number of molecules is more appropriate, in my view, than calculations based on weight (mass) of the poison in the lethal dose. Polonium-210 exists as atoms so using the atomic weight is equivalent to using molecular weights for the other poisons.

The mechanism by which VX agent kills is the same as for other, similar nerve agents known as taubin (GA), sarin (GB), soman (GD), and cyclosarin (GF). All these agents poison the enzyme acetylcholinesterase [22]. Several organo-phosporous compounds act similarly but are far less toxic and some affect insects, not humans. Nerve agents are extremely toxic and act very rapidly. VX agent is designed to be less volatile and to be persistent on materials and the terrain for long periods of time. Human uptake is primarily *via* skin contact, but it can also be inhaled. The route of body entry affects the time to symptoms which include death. Effects occur more quickly when entry is by the respiratory route compared to absorption through the skin. At the biochemical level, VX agent and similar compounds are toxic because they poison an enzyme known as acetylcholinesterase (Fig. **5**). Acetylcholinesterase functions in transmission of nerve impulses. For nerve impulses to activate muscle movement, the electrical impulse must cross a barrier called the myoneural junction. This is an actual, physical gap that the impulse must cross. To do so, a chemical signal is produced by the impulse on the side of the junction nearest the origin of the impulse (CNS side). This chemical signal is acetylcholine. Acetylcholine diffuses across the synapse and stimulates receptors on the muscle innervations site of the junction. The nerve impulse causes muscle contraction. To stop the effect of acetylcholine molecules, they are chemically cleaved by an enzyme called acetyl cholinesterase.

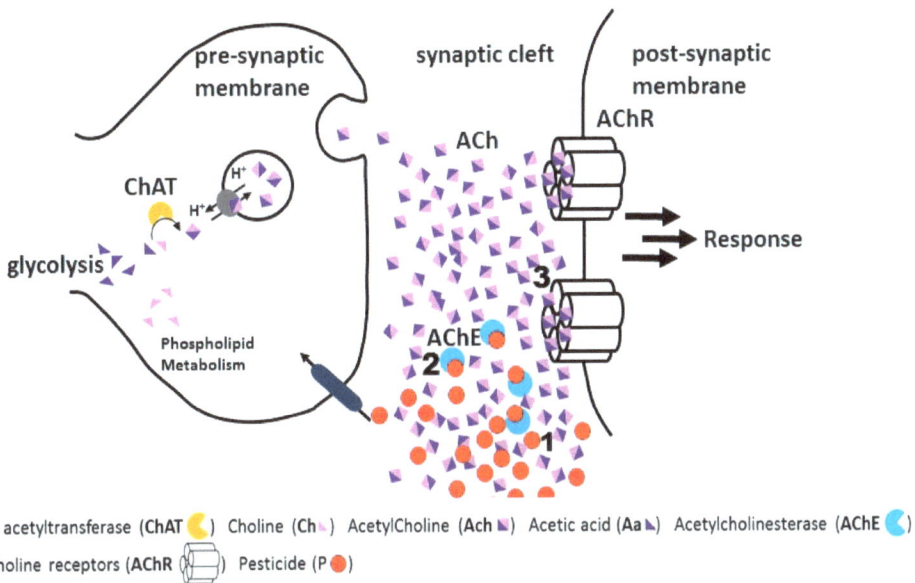

Choline acetyltransferase (ChAT) Choline (Ch) AcetylCholine (Ach) Acetic acid (Aa) Acetylcholinesterase (AChE)
Acetylcholine receptors (AChR) Pesticide (P)

Fig. (5). The myoneural junction site of action of VX; see text for explanation.

The cleavage products are recycled. The important fact is that cholinesterase "erases" the agent that carries the nerve impulse across the synapse to effect muscle contraction. When cholinesterase is poisoned by VX agent, the acetylcholine remains and the nerve continues to function across the synapse and muscle contraction is sustained inappropriately.

Deadly Snake Venom, Taipoxin

Taipoxin is the primary neurotoxin in the venom of a particular Taipan snake, *Oxyuranus microlepidotus*. The average yield of snake venom (when milked) is 44 mg, and the maximum is said to be 110 mg [13]. Taipan snake venom is a complex mixture. Another species, *Oxyuranus scutellatus*, has even more venom: 120 mg average and 400 mg maximum. It is said that the average venom injected at first bite (defensive strike) by *O. Microlepidotus* is 17.3 mg with a range of 0.7 to 45.6 mg. *O. scutellatus* has a neurotrophic venom very similar to *O. microlepidotus*. Both neurotoxins work similarly to botulinum toxin and they block signal sent to muscles and paralysis of the diaphragm impairs breathing and death can occur. More specifically, taipoxin binds to neuromuscular post-synaptic junction sites and prevents nerve impulses from activating muscles which results in a flaccid paralysis. Toxicity of the venom of *Oxyuranus microlepidotus*, reported as toxicity for the whole toxin, is $LD_{50} = 0.025$ mg/kg in mice (18-20 gram) with subcutaneous injection (a referenced source is given) [13]. This site gives several clinical cases including circumstances of envenomation, course of treatment, and outcome; there are 97 total references.

Each snake species has venom that is a complex mixture of proteins and non-proteins that act as neurotoxins, procoagulants, and myolysins. Neurotoxins are present that act both pre-synaptically (taipoxin) and post-synaptically. Coagulants principally are analogues of blood factors involved in coagulation but they act independently of control by cofactors including calcium, and they convert prothrombin to thrombin (to form a clot). Myolysins perform an additional toxic action at presynaptic junctions (this is an additional toxic action of taipoxin). There is some controversy about various snakes that are called taipans, and some claim the inland taipan is the most venomous. A particular snake bite will deliver a variable amount of venom depending on when the snake last used its venom, and the snake is also capable of controlling the amount of venom delivered. Apparently, this is a mechanism that allows the snake to conserve its poison. Some snake bites are described as "dry" to indicate essentially no poison was delivered. It is reported that early symptoms are usually seen in the first six hours, anti-coagulation effects leading to hemorrhaging may develop within 30 minutes; however, systemic collapse, unconsciousness and convulsions may occur (especially in children) occasionally as rapidly as 15 minutes after the bite [13]. For severe envenomation: "… in some cases, paralysis may be sufficiently advanced at a cellular level that antivenom cannot prevent severe paralysis… overall, up to 75% of all taipan bites will prove fatal if no antivenom treatment is used" [13]. An antidote (antitoxin or antivenom; the terms are used interchangeably) is available and with prompt, effective medical treatment, survival is probable. However, antitoxin does not reverse damage already done to nerves, and antivenom therapy may be too late to prevent paralysis and death [13]. The antitoxin is an antibody, usually developed in the horse.

Radioactive Poison

Polonium-210 is an element that is radioactive [23]. An international symbol of radiation hazard has been adopted (Fig. **6**). Emission of radiation is a physical not a chemical action. Does this technically disqualify polonium-210? Did the creator of the definition of poison (see page 1) carefully consider the nuance implied by "chemical action", and does this exclude radioactivity? After all, scientists talk about "radiation poisoning". Think about it. True enough, the emission of an alpha particle is an event involving the atomic nucleus and radiation emission is "physics" and not defined as "chemistry". However, the effect of the alpha particle on a living cell immediately involves chemistry. The toxic event can be a chemical change in a DNA molecule to lead to a fixed mutation, or it can be inactivation of function by physical destruction of cells. So, there is a gray area here that includes an immediate involvement of biochemistry (and thus, chemistry) in the toxic mechanism. I vote to include polonium as a poison.

Fig. (6). International symbol of radiation hazard.

There are, however, special complexities encountered when considering radiation. All chemical poisons must be absorbed into the body, and transported and distributed to the site of action where further chemistry must occur. A poison that is toxic by chemical action, if it is in a container or otherwise outside the body, is not a hazard to human life. A molecule of chemical poison, even though ingested, has no deadly effect if it is purged before absorption. A molecule of chemical poison though ingested or injected has no deadly effect if it is blocked (inactivated) before it reaches its cellular site of action. A molecule of chemical poison on the skin is ineffective if it is removed before penetration and absorption. Radioactive atoms obey these rules to an extent. However, radioactivity implies generation of particles or rays which can have a variety of properties in terms of: penetrability, rate of production, and interaction with biological tissue (Fig. 7). Both polonium-210 and radioactive carbon (C-14), for example, emit alpha particles and they are not a significant hazard unless consumed. Radioactive compounds that emit beta particles or gamma rays (which possess different energy levels) can be deadly from sources outside the body.

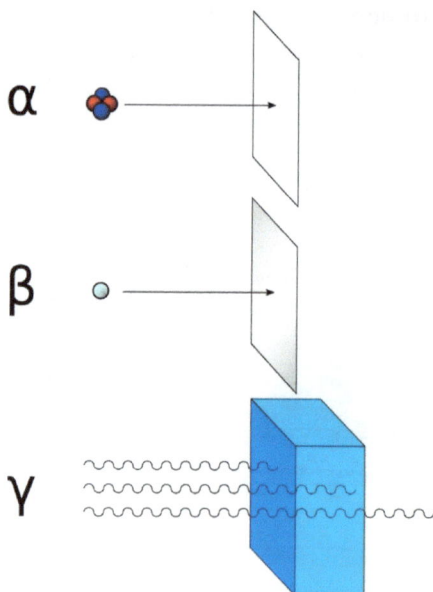

Fig. (7). Alpha particles, emitted by polonium-210, are blocked by a simple sheet of paper; beta particles and gamma rays are much more penetrating and some gamma rays pass through a considerable thickness of lead.

The atomic weight of polonium-210 (relevant for this comparison to the molecular weights of the other poisons) is 267 grams per gram atomic weight and its lethal dose is stated to be 0.89 micrograms for an 80 kilogram person [4]. This is equivalent to 0.78 micrograms for a 70 kilogram person which is the reference weight used in comparative toxicology. Compared to botulinum, it takes 11.1 times as much to be lethal (Table **1**). Because this element's toxicity is caused by its radioactivity, its ability to poison people has some unusual characteristics. The radiation emitted is primarily alpha particles and these particles penetrate poorly, including that they are blocked by a sheet of ordinary paper and also by the skin (Fig. 7). Thus, polonium-210 must be ingested or injected to be deadly. Because radiation poisons the body by mechanisms that are quite different from the actions of the other chemicals in Table **1**, the calculation of its toxicity is complicated and details are provided near the end of the chapter. Radiation poisoning with the minimum lethal dose of polonium-210 is a slow process; there is no effective antidote and death from ingesting a lethal dose of radiation is practically certain.

Table 1. The four most toxic substances, assessed by four relevant factors.

Poison	Lethal Dose[a] LD_{50}	Lethal Dose[b] LD_{50}*	Death Time (hours)[c]	Antidote[d]
Botulinum A	1×10^{-6}	4.0×10^{9}	24	Yes
Taipoxin	2.0	2.6×10^{16}	1	Yes
Polonium	11.1×10^{-6}	3.2×10^{13}	1,440	No
VX agent	0.143	3.2×10^{17}	0.1	Yes-No

[a] The traditional ranking of the toxicities of poisons is by the LD_{50}, the amount in mg of toxin per kilogram of body weight of the test subject that causes death in half the subjects. Further information is provided in the text.
[b] Poisons, in the opinion of the author, should be ranked based on the number of molecules that cause death. There are 6.02×10^{23} molecules in one gram molecular weight of a substance (one gram atomic weight for polonium). Note the exponents. Further explanation is provided in the text.
[c] Time to death is very difficult to assess and will be discussed in the text.
[d] Antidotes are available for these poisons except for polonium-210. However, except under certain battlefield conditions, described in the text, there is insufficient time (because death occurs rapidly) to administer the antidote for VX agent.

REEVALUATING THE MOST TOXIC POISONS

Certainly, by the traditional meaning of poisons, botulinum toxin wins as the world's most toxic substance. This is valid based on the amount required to kill. However, it is also possible to consider the time required to kill and whether one can be saved by an antidote (Table **1**). There is, however, another way to evaluate what is the deadliest poison. Suppose we keep the dictionary definition of poison (a substance that by chemical action usually kills, injures or impairs an organism) [1], but extend our thinking to include things that are themselves alive, or at least have the ability to replicate (increase in number) when introduced into the human body. Perhaps you protest: these are called infectious agents, not poisons. Think of botulinum toxin. It is only a complex chemical produced by a germ during its growth. The poison is produced (usually), in a jar or can of improperly canned food; or in modern times, in a biological laboratory devoted to germ warfare. The toxin accumulates in the food and the disease is called a food "intoxication" to distinguish it from a food "infection". The germ, itself is not the poison; however, the germ is the necessary agent for the poison.

If we consider the germ as the poison, we can see how this affects our calculation. Indeed, is it possible for the botulism germ to grow inside a person and poison that person? Indeed, adult intestinal, toxemia botulism and wound infection botulism are known [10]. How does this affect our evaluation? How much does one botulinum cell weigh? We also have to assume that a single cell could be infectious. My evaluation is that a single cell of *Clostridium botulinum* weighs approximately 1 picogram (1 x 10^{-12} gram), and this is in general agreement with other data [24]. The somewhat astonishing calculated result is that the lethal dose, if we assume it to arise from a single cell, is 4 x 10^4 (40,000) times the weight of a single botulinum cell. This implies that the lethal dose of toxin is produced by the action of many botulism cells. It is also consistent with the fact that the infectious dose is not one cell, and that the lethal dose of toxin is produced over time. To calculate the time required for one cell to produce the lethal dose of toxin, we must know the rate of toxin production. Since the botulinum cells are multiplying during this interval of toxin production, the doubling time of the cells is also a relevant factor.

There is information about rate of production of botulinum toxin. In a study published in 1979 [25] in a fermenter under controlled conditions, the maximum concentration of type A botulinum produced was 6.3 x 10^5 mouse medium lethal doses per ml and "was attained within 24 hours". Growth of the cells was exponential for about 6 hours with a mean generation time of 76 minutes, but continued significantly until 24 hours and the number of cells remained stable through 72 hours. They concluded that cell lysis was not required to obtain maximum toxin concentration, implying that the toxin is excreted by the cells. Derise *et al*. [26] gives details in 2013 about production and purification of botulinum type A and defines the Unit used in therapy. This Unit, however is not universal [27]. My quest was to determine the amount of botulinum toxin-A produced on a per cell basis. This proved to be impossible due to lack of data about the number of cells per ml of culture [25, 26].

However, both papers agree that it is reasonable that 7.56 x 10^{-4} grams per milliliter of botulinum-A was produced in cultures under the optimum conditions of the tests. Using the fact that 1.2 ng is the LD-50 for the mouse (very similar to the value used in Table **1**), this is equivalent to 630,000 mouse lethal doses per milliliter of culture. In terms of lethal doses for one human, this is 9,000 per ml. This is the worst-case scenario for a can of spoiled food; the smallest taste could be lethal!

Let us go even further in stretching the concept of poison. The immune system of humans can react to some substances, called allergens, with a potentially lethal reaction known as anaphylaxis. I know of no toxicology assessment that includes agents that are toxic by this means. Many substances, including dangerous poisons, also can induce anaphylaxis in susceptible people. However, the individual must first be made susceptible by an inducing dose and passage of sufficient time for the allergic mechanism to develop. Therefore, the toxicity is not direct, and it is understandable that pathology by this mechanism is not usually called toxicity.

There were 2,458 anaphylaxis-related deaths in United States from 1999 to 2010. Medications were the most common cause (58.8%), followed by unspecified cause (19.3%), venom (15.2%), and food (6.7%) [28]. The authors state that the number of venom-caused anaphylaxis was probably underestimated. Anaphylaxis is surely the cause of death in some envenomations. Assessing a lethal dose for an anaphylaxis agent is most difficult. One reference states that the median lethal dose (LD-50) for bee venom is 2.8 mg per kg of body weight which is 196 mg [29]. This is equivalent to 196,000 micrograms of bee venom for a 70 kg person (ordinarily the standard for toxicological comparisons of poisons). Assuming all the bee venom is injected (0.3 mg venom per sting) 653 bee stings would be lethal for a 70 kg person. Honeybee venom is reported (based on cited references) to be a transparent, odorless, bitter-tasting liquid that causes burning and irritation to the human eye that contains a number of very volatile compounds and is rich in enzymes, peptides and biogenic amines [29]. This reference provides tabulated details about bee venom composition and descriptions of allergic reactions that range from non-allergenic at the time of the sting, non-allergenic days after the sting, large local reaction, cutaneous allergic reaction, non life-threatening reaction, life-threatening systemic allergic reaction with unconsciousness, hypotension or fainting, and respiratory distress and laryngeal blockage from massive swelling in the throat (the latter is anaphylaxis) [29]. A comparison of LD-50 values (reported as mg/kg for mice) for venoms from various *Hymenoptera* are as follows: Honey bee 2.8, velvet ant 71, paper wasp 2.4, yellow jacket 3.5, and two values are given for harvester ant: 0.66 and 0.12 (references for these values are cited) [30].

Peanut is an example of a substance that can kill sensitive people in minutes by anaphylaxis. The more usual allergic reaction is called the minimum eliciting dose (ED) and modified food challenge protocols have been used to assess this dose. Alan Khadavi reported results from a study of 63 children with peanut allergy, with a design that attempted to model real life exposure [31]. Children received, at two hour intervals, from 3 mg to 400 mg of peanut protein until a reaction occurred (the average peanut was said to weigh about 1,000 mg). Within 45 minutes 45 children had an allergic reaction; some occurred within 30 minutes; and most within 1 hour. The ED was usually less than 2 mg of peanut protein. The author concluded that it was not safe for children to consume certain food which were considered to be peanut free if the food was processed in a plant that processed or cooked peanuts [31].

A review of peanut allergy was authored by Saleh Al-Muhsen, Ann E. Clark and Rhoda S. Kagen in 2003 [32]. They state that: "Peanut allergy accounts for the majority of severe food-related allergic reactions. It tends to present early in life, and affected individuals generally do not outgrow it. In highly sensitized people, trace quantities can induce an allergic reaction". They state that: "One peanut contains about 200 mg of protein (and provide a reference). In most people with peanut allergy, symptoms develop after substantially less than 1 peanut is ingested, and highly allergic people can react to trace amounts. In one study designed to determine the minimum dose of peanut protein capable of eliciting an allergic reaction in highly sensitized individuals, subjective symptoms were reported with doses as low as 100 micrograms, and objective signs were evident at 2 mg (reference provided). They also state that in more than 70% of children with peanut allergy, symptoms develop at their first known exposure (two references are provided), and conclude that occult exposure must have occurred. It is known that the peanut allergy response is IgE-mediated; therefore, a sensitizing dose must precede the eliciting dose.

For our purposes, neither the data for peanuts or insect venom is very satisfactory because lethal doses are not known. These results are included to verify that a commonly experienced type of envenomation has not been overlooked. Because of the poor quality of the data available, the complexity of the association (through anaphylaxis after a dose that establishes allergy), and the indications that a comparatively large dose is required, I conclude that agents that cause allergic responses do not qualify as the most toxic of substances.

SCIENTIFIC AND MEDICAL DETAILS AND CALCULATIONS

Overview

Calculation of toxicity is mathematically complex. As examples I will list the following. Most reports for human lethal doses of toxins are based on data from exposures of rats, mice or other animals. Experimental laboratory conditions vary significantly including whether the toxin is injected, ingested, inhaled, or occurs by skin contact. The following is an example of the complexity of quantifying the effects of poisons. I calculated the lethal dose of VX agent to be 900 micrograms (for inhalation of vapor) from publication listed as authored by the "Federation of American Scientists, dated 2013 [6]; or 1218 micrograms based on my calculation from the reported 17.4 micrograms per kilogram using adult swine [33]; or 4.35 micrograms by another reference [34]. I compromised by using a figure of 10 mg for percutaneous application because converting inhalation dose (although they are useful for many purposes) to injected dose is difficult because the latter involves a time period of inhalation (and is not equivalent to a single dose) and other complications. It is agreed, also, that skin application is not equivalent to injection.

Poisoning by Polonium

Polonium-210 requires the most time to kill of the highly toxic substances (Table **1**). The fact that there is no antidote for polonium-210, in my opinion, is important in considering its ranking in our list of toxic substances. However, it is difficult to assess the consequences of antidotes for several reasons. To affect the outcome of a poisoning, medical assistance must be available, a competent diagnosis must be made, and the antidote must be available and competently administered in a timely manner. After a lethal dose of Polonium-210, death is unlikely before 30 days and no antidote or treatment (other than palliative) is available.

The lethal amounts listed for polonium-210 differ significantly in three references: "probably on the order of 10-30 micrograms" [15] (based on provided references) [the] "estimated acute lethal dose from oral ingestion for an adult if untreated is 10-30 micrograms, and as little as 1 microgram might be lethal to the most radiosensitive members of the population" [35], and 0.89 micrograms for an 80 kg person [4]. Approximately 10 micrograms was said to have been used in the "infamous" poisoning death of Alexander Litvinenko [36]. This calculates to be more than 11-times the lethal dose; both numbers on which the estimate is based are surely only approximations.

My proposal that poisons should be ranked based on lethal dose calculated as the number of molecules affects assessment of the toxicity of polonium-210. This affect starts with the fact that radioactive polonium-210 acts as an atom, not a molecule. Because poisonous molecules vary greatly in their size (mass or weight) this greatly influences the results. It is most reasonable to rate poisons this way because at the deepest mechanism level poisons affect molecules that are parts of the structure and metabolism of cells. The way most of these poisons work at the cellular and even the molecular level is reasonably known. All poisons ultimately are toxic because they damage or interfere with some structure or function to a significant extent and cause the destruction of a function vital to life. Polonium-210 emits radiation which damages most chemicals that the radiation happens to strike. This damage is cumulative but extremely varied in its target. Some damage is relatively unimportant to the cell's well being; some is important long term by causing mutations and some damage cascades into non-survivable consequences. As stated, the radiation is alpha particles and these are rapidly absorbed and will not penetrate a sheet of paper which is adequate shielding. However, polonium-210 is highly radioactive with a specific activity of

166 terabecquerel per gram. A terabecquerel is 10^{12} becquerels, 1 Curie is $3.7x10^{10}$ becquerels, and 1 curie is 3.7×10^{10} disintegrations per second [37]. Thus, a single microgram (one-millionths of a gram) of polonium-210 has 166 million disintegrations per second, each of which produces an alpha particle. The radioactive half-life is 138 days [37]; however its effective half-life in the body is about 30 days because of excretion [4]. The alpha particles deposit their energy in a small volume of tissue (they do not penetrate far). Alpha particles can disrupt cell structure, damage internal sub-cellular components including the nucleus, damage DNA itself, and cause other pathology that can result in cell death [37]. With a minimal lethal dose no effects are said to be expected for about a week, mild symptoms occurring over the following week, and serious illness and/or death is expected in approximately a month [4].

Poisoning by Nerve Gases

The complexity of assessing poisons is exemplified by the voluminous report by a subcommittee of scientists appointed by the National Research Council of the National Academies of Science [34]. This report states that the human toxicity of the VX agent could not be established because there was insufficient laboratory animal data to derive a dose that meets the criteria for AEGL-3. AEGL means "acute exposure guideline level" and 3 indicates it is the level for: "The airborne concentration (expressed as parts per million or mg/M^3) of a substance above which it is predicted that the general population, including susceptible individuals, could experience life-threatening health effects or death" [34]. The purpose of these guidelines is to assess safety of exposures to small doses of hazardous substances. Clearly, this is not the same as the commonly used definition for the lethal dose for 50% of test subjects (LD_{50}).

VX agent kills very quickly by reacting chemically with a substance essential for transmission of nerve impulses across gaps called synapses. Stopping nerve transmission necessary to activate breathing is rapidly lethal because there are a relatively small number of such synapses compared to the number of molecules of poison in even a small dose.

VX agent is successfully reversible by the antidotes atropine and alprazolam. The antidotes must be applied within the first minutes after exposure and this is unlikely to happen except on the battlefield in circumstances where soldiers are trained to use the "buddy system"; are knowledgeable about the symptoms of VX; and have the antidotes as part of their equipment.

It is readily appreciated that human experiments cannot be done to determine the LD_{50} for nerve agents; indeed, for humane reasons, the LD_{50} (even with laboratory animals) is no longer (or rarely) tested today. The committee [34] found there was insufficient data for VX agent, but did find sufficient data for the related substance GB (sarin) and they found it "...was possible to develop AEGL estimates for agent VX agent by a comparative method of relative potency analysis from the more complete data set for nerve agent GB." They noted that this method had previously been used in peer reviewed reports. Additionally, toxicity was assessed based on effects of GB on cholinesterase, the enzyme site of the toxicity of the nerve agents. Data from human cholinesterase (the site of action of the poison) consistently indicated that the GB to VX agent relative potency ratio was 4 (VX agent was four times as potent as GB). This report [34] further stated that the "... AEGL-3 values for agent VX agent were derived from recent inhalation studies in which the lethality of GB to female Sprague-Dawley rats was evaluated for the 10-, 30-, 60-, 90-, 240-, and 360-min time periods (the authors provided a reference for this)". It should be noted that, as is customary in such risk assessments, factors are applied to the data to convert from rat toxicity to human toxicity, to account for the most susceptible individuals in a population, and for the fact that the data sets were sparse. Subsequently, a total uncertainty factor of 100 was used in calculating the AEGL-3 results.

This means, that because of data uncertainty and because the reported information is designed to be protective, the AEGL-3 dose is made smaller by a factor of 100. The reported AEGL-3 number for 10 minutes of exposure to VX agent is 0.029 mg per m^3. Assuming a respiration rate-volume of 15 liters per minute, and doing the math, this interpolated lethal dose for a human is 4.35 micrograms. As stated previously, this includes a reduction by a factor of 100 (to provide safety for the intended use of AEGL-3). Thus, for a LD_{50} comparison, the value is 435 micrograms. Indeed, another report [6] gives a value for VX agent of 30 mg min/m^3 based on a respiration rate of 15 liters per minute and exposure of 2 to 10 minutes. This calculates to 450 micrograms as the lethal dose (assuming the values in [6] are based on 10 minutes (the report states estimated exposure durations were 2-10 minutes). If we increase this by the factor of 100 (the uncertainty factor in the AEGL-3 value), the comparison is 435 micrograms to 450 micrograms. This is reasonable (even remarkable) agreement. The reader may find the results unsatisfactory, however. It is the best that can be done to make sense of the complex data. Therefore, I used the value 450 micrograms (Table **1**).

The outcomes using my system are very dependent on the effect ascribed for antidotes. For polonium there is no antidote and, therefore, no argument. For the other poisons for purposes of this chart, I make the assumption that competent medical assistance is available but recognize that except under the battle field circumstances previously described, VX agent is so rapidly acting that antidote is extremely unlikely to be used successfully.

TOXICITIES OF SELECTED POISONS

As previously discussed there are many way to compare the toxicities of agents. The most-used method is what is called the lethal dose for 50% of test subjects (LD_{50}). The dose is measured in mg of poison per kg body weight of the test species, with the species identified. Sometimes it is even necessary to specify beyond the species to strain or other identifier. It is essential to identify the route of poisoning such as intraperitoneal injection, intradermal injection, orally, or *via* inhalation, for example. It often is of most interest to know the effect of the poison on the human. Of course, it is not possible to conduct such experiments. For some poisons, data may be available because of some accidental circumstances, but such data is usually only for a single individual. Of course, there are individual differences which can be quite large, in the amount of a poison required to kill. Therefore in laboratory experiments, for many years, it was customary to test various doses of a poison given by various routes on a significant number of animals all of the same species, within a narrow age and weigh range and all of the same sex. Diet, light and dark cycle, and temperature in the cage environment were controlled. From data obtained by observation it was possible to calculate the dose that would kill with a certain statistical probability a specified proportion (%) of the animals. The LD_{50} is often published as the most useful value. The amount of a given poison required to kill all of the test animals in a group is a much larger value and is also related to the number of animals in the test group. The LD_{50} is a much more reproducible figure and is more representative of the dose required to kill an "average" member of the species.

An additional problem arises when the focus is to determine the dose for a human. It often is useful to compare the dose measured from a laboratory study using the mouse or the rat and some factor to convert the results for humans. The simplest idea was to devise a factor that represented the ratio of the weight of the test animal to a human and multiply the LD_{50} for the animal by that number to arrive at the LD_{50} for the human. This has flaws because it is wrong to assume that poison on weight differences for individuals, would also scale across species lines where weight differences, organ size, enzyme content, and many other relevant features do not scale equally. Therefore, after much research, it has become customary to use a derived conversion factor called a "scaling" factor and such factors are generally agreed on and can be found in published sources such as: 'Guidance for Industry' [38].

Traditional LD_{50} Toxicities

The toxicity of agents is most often presented as the LD_{50} as previously described. Because the human toxic dose is often of primary interest, the conversion of data from the mouse or other laboratory animal is done by a formula to determine the human equivalent dose (HED) and this will be described subsequently.

For botulinum toxin, type A is acknowledged to be the most toxic. The best data I have found is that of Michael Gill who reported the LD_{50} is approximately 1 ng per kg body weight (0.000,001 mg/kg). This is based on values reported for mice, guinea pigs, rabbits, and monkeys and the range for all these species was only from 0.5 to 0.7 ng for monkeys to 1.2 ng for mice. Gill cites a source that states: "Humans are said to be at least as sensitive as mice". Stephen Arnon, *et al.*, makes similar conclusions about the toxicity values and provides the molecular weight as 150,000 Da. with a heavy chain of 100,000 and a light chain of 50,000. I advise extreme caution in accepting published values; I have found various other values in publications, including errors in extrapolating doses.

Proposed LD_{50*} Toxicities

I am proposing that an additional way of reporting the toxicities of agents is appropriate and useful. It is called the LD_{50*} (Table **2**). Toxic molecules come in a great range of sizes based on differences in their molecular weights (masses). The science of poisons, based on the current sophistication of science, is focused on mechanisms. Indeed, the way chemicals cause toxicity is known for most toxins at the molecular level. Since toxicity always occurs molecule-by-molecule, the truism that 'the dose makes the poison' is most true at the molecular level. Size matters. It continues to be useful to report toxicity in grams of the toxin as has been done traditionally. I propose that it is also useful to compare toxicities based on the number of molecules that cause death ($LD_{50}*$).

It is a simple matter to convert LD_{50} to $LD_{50}*$. The molecular weight of the toxin must be known and expressed in grams. The process is: convert the dose (as usually expressed) from mg/kg to g/kg (divide by 1,000). Multiply the results by Avogadro's number (6.02×10^{23}) and divide by the molecular weight of the toxin in grams (Da units).

Another way of saying this is: the LD_{50} (based on mg of toxin) is converted to the new unit $LD_{50}*$ (based on number of molecules) by a factor equivalent to the dose in grams divided by the molecular weight of the toxin in grams multiplied by Avogadro's number (6.02×10^{23}).

There are complications in interpreting both the traditional $LD_{50,}$ and the $LD_{50}*$ that I propose. For the $LD_{50}*$ the assumption is that the poison is toxic based on its action as administered, on a molecular basis. This is a good assumption for many toxins. However, some toxins act not as the molecules delivered but as a metabolized species, which may be different in molecular weight. In some cases, the toxic species may simply be an ionized fraction of the delivered dose. This is true for cyanide. However, even for potassium cyanide, the mole fraction that is the toxin (the cyanide ion) is the same as obtained by using the weighed dose.

To convert the dose from a species of laboratory test animal to the human equivalent dose, a customary scaling factor is used. An accepted table of conversion factors for various laboratory species is available in a 'Guidance for Industry' document [38]. This calculated value is in mg/kg body weight and to get the total dose for a human, the value must be multiplied by the weight of the human in Kg; 70 kg is normally used for a human. In the cited document [38], the conversion factor for converting mouse dose to human equivalent dose (mg/kg) is 0.081, and for the rat 0.162. These factors are not based on the weight ratios of the lab animal species and the human, but on a factor based on body surface area. For example the ratio of body weights of man and mouse is approximately 70 kg/0.02 kg = 3,500 and this value is not useful for directly comparing doses in mg/kg between the species. A better comparison than body weight is body surface area. Body surface area comparison are provided in a table in the cited source [38]. These data can be used to compare doses across species (generally from a laboratory species to the human). For example, to convert dose in mg/kg body weight to dose in mg/m^2 body surface area, the ratio of these values (7/37, see table in reference for explanation) is multiplied by the dose for the rat in mg/kg.

To calculate the human equivalent dose (HED), when the rat dose is 75 mg/kg, the equivalent dose in the human is 75 divided by 7/37 which equals 14 mg/kg. An excellent tutorial on key principles of toxicology is available on-line [54].

Sources for LD_{50} Values

LD_{50} values are widely available for most every chemical Table **2**. However, finding reliable values is more difficult. It is desirable to have a referenced source that also provides details of the tests that are the basis of the reported value. The following are suggestions with some information about the sources. All commercially available chemicals will have a material data sheet (MSDS) that is

Table 2. Toxicities of agents with the LD_{50} converted to LD_{50*} [from weight (mass) of poison to number of molecules of poison (rat oral dose): see text for details].

Toxic Agent [a]	LD_{50*}	LD_{50} Reference
Water	3.01×10^{24}	[39]
Methanol	1.1×10^{23}	[40]
Ethanol	9.0×10^{22}	[41]
Glucose	8.6×10^{22}	[42]

(Table 2) contd.....

Toxic Agent [a]	LD$_{50*}$	LD$_{50}$ Reference
Sucrose	5.2×10^{22}	[42]
Sodium Chloride	3.0×10^{22}	[43]
Morphine Sulfate	2.1×10^{21}	[44]
Ibuprofen	1.9×10^{21}	[45]
Coumarin	1.2×10^{21}	[46]
Doxorubicin [b]	7.3×10^{20}	[47]
Aspirin	6.7×10^{20}	[48]
Caffeine	5.9×10^{20}	[49]
Paraquat	4.8×10^{20}	[50]
Arsenic Trioxide	4.3×10^{19}	[51]
Fentanyl [c]	5.6×10^{18}	[52]
Brodifacoum [c]	3.2×10^{17}	[53]
Botulinum A [d]	4.0×10^{9}	[11]

[a] Toxicity is strongly influenced by route of administration and species tested.
[b] Mouse (not rat) oral 698 mg/kg; i.p. 11.2 mg/kg; and rat i.v. 13.1 mg/kg.
[c] Estimated lethal dose in humans is 2 mg (0.029 mg/kg).
[d] Human, best estimate based on several sources cited in the reference given, rats and mice similar.

available without cost and usually on-line. The toxicity data in MSDSs are generally reliable.

For most toxic agents there are many published scientific papers and the key to recovering those with a LD$_{50}$ value is to look for older publication. Testing lethal doses of chemicals on rats and mice is discouraged today. The title of a review published in 1982 suggests the scope of a useful publication: "Bacterial Toxins: a Table of Lethal Amounts" [11]. The website "Mechanisms of Acute Toxicity" by Dan Wilson at Dow Chemical Company is informative but does not provide references [55]. This site compares acute classification schemes as authorized by the EPA, SPSC, OSHA, DOT, and OECD. For example, the EPA has four categories for classifications of chemicals based on toxicity (LD$_{50}$ mg/kg). Toxicity Class I: < 50, most toxic, requires signal word 'Danger-Poison' with skull and crossbones symbol; Class II: 50 to 500, moderately toxic, signal word: 'Warning'; Class III: 500-5,000, slightly toxic, signal word 'Caution'; Class IV: >5,000, practically nontoxic, no signal word required.

The website 'Venom Supplies' has LD$_{50}$ values for a large number of venoms, and provides a useful 'relative toxicity' scale [56]. For example it compares Australian snake venoms: Inland taipan 50.0, common brown snake 12.5, taipan 7.8, copperhead 1.0, (other snakes are also listed). It also compares some non-

Australian snakes: Indian cobra 1.0, King cobra 0.3, Eastern diamond-back rattlesnake much less than 0.1.

A website by the Oxford Treatment Center is useful for opioid drugs [57]. The site says: "The following amounts are generally considered to be lethal doses for these opiate drugs for an average person who has not developed tolerance: Morphine doses of over 200 mg are considered to be lethal... The lethal dose for heroin is generally reported as being between 75 and 375 mg... The lethal dose for hydrocodone is generally stated to be around 90 mg...A single dose of 40 mg or more of oxycodone may produce lethal effects in some individuals... The lethal dose for fentanyl is generally stated to be 2 milligrams.... Based on the above figures, one can calculate that *lethal dose for fentanyl is approximately 100 times less than the lethal dose for morphine... The lethal effects that occur as a result of fentanyl overdose are most often due to significant respiratory suppression or the complete halting of breathing as a result of the central nervous system depressant effects of the drug.*" (Italics are in the original.) The title of another site that has LD_{50} values gives a good summary of its scope: 'The Biological and Toxic Weapons Threat to the United States" [58]. A published paper gives lethal dose for mice of Australian and other snake venoms [59]. A technical report of the U.S. Army Center for Environmental Health Research, Fort Detrick, MD provides derived human lethal doses [60]. Representative LD_{50} values were complied for 48 toxins; however, reference sources are not provided [61]. A website [62] lists the LD_{50} for many biological toxins, provides species tested and route of administration but unfortunately does not provide reference sources. It should be kept in mind that there is a degree of imprecision in LD_{50} data and although it is extremely useful, different values are reported and result from differences in the details of testing. I hope that the concept of the LD_{50*} based on the number of molecules in a lethal dose will become useful.

NOTES

[1] Theophrast Paracelcus: *Die drittle Defension wegen des Schreibens der neuen Rezptem Defensiones 1538. Werke Bd. 2, Darmstadt 1965, p. 510.*

REFERENCES

[1]   Poison [Internet]. Merriam-Webster Disctionary. [cited 2016 Jan 1]. Available from: https://www.merriam-webster.com/dictionary/poison

[2]   Arnon SS, Schechter R, Inglesby TV, Henderson DA, Bartlett JG, Ascher MS, *et al.* Botulinum Toxin as a Biological Weapon. JAMA [Internet] American Medical Association 2001 Feb 28[cited 2017 Apr 29]; 285(8): 1059. Available from: http://jama.jamanetwork.com/article.aspx?doi=10.1001/jama.285.8.1059
[http://dx.doi.org/10.1001/jama.285.8.1059]

[3]   Sobel J, Tucker N, Sulka A, McLaughlin J, Maslanka S. Foodborne botulism in the United States, 1990-2000. Emerg Infect Dis 2004; 10(9): 1606-11.

[http://dx.doi.org/10.3201/eid1009.030745] [PMID: 15498163]

[4]   Sublette C. Polonium Poisoning [Internet] , [cited 2017 Apr 27]; Available from: http://nuclear weaponarchive.org/News/PoloniumPoison.html

[5]   9 of the World's Deadliest Snakes | Britannicacom [Internet] , [cited 2016 Dec 22]; Available from: https://www.britannica.com/list/9-of-the-worlds-deadliest-snakes

[6]   Types of Chemical Weapons [Internet] , [cited 2016 Dec 22]; Available from: https://fas.org/ programs/bio/chemweapons/cwagents.html

[7]   The coward's weapon, poison - John Fletcher - BrainyQuote [Internet] , [cited 2017 May 8]; Available from: https://www.brainyquote.com/quotes/quotes/j/johnfletch177102.html

[8]   John Fletcher | English dramatist | Britannicacom [Internet] , [cited 2017 May 8]; Available from: https://www.britannica.com/biography/John-Fletcher

[9]   History of the Skull & Crossbones Symbol Used in Poison Warning Signs and Labels [Internet] , [cited 2017 May 4]; Available from: http://www.mysafetysign.com/poison-symbol-history

[10]   Sobel J. Botulism. Clin Infect Dis [Internet]. 2005 Oct 15; [cited 2016 Dec 22]; Available from: http://www.ncbi.nlm.nih.gov/pubmed/16163636

[11]   Gill DM. Bacterial Toxins: a Table of Lethal Amounts. Microbiol Rev [Internet] American Society for Microbiology (ASM) , 1982 Mar; [cited 2017 Apr 29];46(1): 86-94. Available from: https://www.ncbi.nlm.nih.gov/pmc/articles/PMC373212/pdf/microrev00066-0096.pdf

[12]   Representative LD. Representative LD 50 Values [Internet] , [cited 2017 Jan 1]; Available from: http://biology.unm.edu/toolson/biotox/representative_LD50_values.pdf

[13]   White J, Covacevich J. Oxyuranus microlepidotus [Internet] , [cited 2017 May 1]; Available from: http://www.inchem.org/documents/pims/animal/taipan.htm

[14]   Environmental 210 Po and its low-level effects 2012.

[15]   Jefferson RD, Goans RE, Blain PG, Thomas SHL. Diagnosis and treatment of polonium poisoning. Clin Toxicol [Internet] Taylor & Francis , 2009 Jun 3; [cited 2016 Dec 22];47(5): 379-92. Available from: http://www.tandfonline.com/doi/full/10.1080/15563650902956431 [http://dx.doi.org/10.1080/15563650902956431]

[16]   Cendron L, Mičetić I, Polverino de Laureto P, Paoli M. Structural analysis of trimeric phospholipase A2 neurotoxin from the Australian taipan snake venom. FEBS J 2012; 279(17): 3121-35. [http://dx.doi.org/10.1111/j.1742-4658.2012.08691.x] [PMID: 22776098]

[17]   Knox JN, Brown WP, Spero L. Molecular Weight of Type A Botulinum Toxin , 1970 [cited 2017 Apr 27];1(2): 205-6. Available from: https://www.ncbi.nlm.nih.gov/pmc/articles/PMC415879/pdf/iai 00290-0067.pdf

[18]   DasGupta BR, Rasmussen S. Purification and amino acid composition of type E botulinum neurotoxin. Toxicon [Internet] , 1983 Jan; [cited 2017 May 4];21(4): 535-45. Available from: http://linkinghub. elsevier.com/retrieve/pii/0041010183901319 [http://dx.doi.org/10.1016/0041-0101(83)90131-9]

[19]   Kedlaya D. Botulinum Toxin [Internet] , [cited 2017 Apr 27]; Available from: http://emedicine. medscape.com/article/325451-overview

[20]   The smallest protein | Science 20 [Internet] , [cited 2017 May 4]; Available from: http://www.science20.com/princerain/blog/smallest_protein

[21]   FDA approves first botulism antitoxin for use in neutralizing all seven known botulinum nerve toxin serotypes [Internet] FDA News Release 2013. Available from: http://www.fda.gov/ NewsEvents/Newsroom/PressAnnouncements/ucm345128.htm

[22]   Nerve agents: introduction, physical and chemical properties, binary technology, mechanism of action, symptoms, antidotes and methods of treatment [Internet] , [cited 2017 Apr 29]; Available from:

https://www.opcw.org/about-chemical-weapons/types-of-chemical-agent/nerve-agents/

[23]   MacGill M. Polonium-210: Why is Po-210 So Poisonous? [Internet] , [cited 2016 Jan 1]; Available from: http://www.medicalnewstoday.com/articles/58088.php

[24]   Mass of a Bacterium - The Physics Factbook [Internet] , [cited 2017 May 4]; Available from: http://hypertextbook.com/facts/2003/LouisSiu.shtml

[25]   Siegel LS, Metzger JF. Toxin production by Clostridium botulinum type A under various fermentation conditions. Appl Environ Microbiol [Internet] American Society for Microbiology (ASM) , 1979 Oct; [cited 2017 May 7];38(4): 606-11. Available from: http://www.ncbi.nlm.nih.gov/pubmed/44175

[26]   Derise N, Harrison K, Juneau C, Rees D. Production of Botulinum Toxin , 2013 [cited 2017 May 7]; Available from: http://nderise.weebly.com/uploads/2/4/4/1/24410337/ be_3340_process_report_final-_group_5.pdf

[27]   Harper L. Botulinum Toxin A—When is a Unit Not a Unit? J Urol [Internet] , 2009 Jan; [cited 2017 May 7];181(1): 414-5. Available from: http://www.ncbi.nlm.nih.gov/pubmed/19019386

[28]   Jerschow E, Lin RY, Scaperotti MM, McGinn AP. Fatal anaphylaxis in the United States, 1999-2010: Temporal patterns and demographic associations. Allergy Clin Immunol [Internet] , 2014 Dec; [cited 2017 May 7];134(6): 1318-1328.e7. Available from: http://linkinghub.elsevier.com/retrieve/ pii/S0091674914011907

[29]   Al-Samie MA, Ali M. Studies on Bee Venom and Its Medical Uses. Int J Adv Res Technol [Internet] , 2012 [cited 2017 May 7]; Available from: http://www.ijoart.org/docs/ Studies-on-Bee-Venom-an--Its-Medical-Uses.pdf

[30]   Which Insect Has the Most Toxic Venom? [Internet] , [cited 2017 May 7]; Available from: https://www.thoughtco.com/which-insect-has-the-most-toxic-venom-1968411

[31]   Khadavi A. Peanut Allergy Eliciting Dose [Internet] , [cited 2017 May 7]; Available from: http://allergylosangeles.com/allergy-blog/much-peanut-will-cause-allergic-reaction/

[32]   Al-Muhsen S, Clarke AE, Kagan RS. Peanut allergy: an overview. CMAJ [Internet]. Canadian Medical Association , 2003 May 13; [cited 2017 May 7];168(10): 1279-85. Available from: http://www.ncbi.nlm.nih.gov/pubmed/12743075

[33]   Langston JL, Myers TM. VX toxicity in the Göttingen minipig , 2016 [cited 2017 May 2]; Available from:
http://ac.els-cdn.com.proxy.mul.missouri.edu/S0378427416332854/1-s2.0-S0378427416332854-main. pdf?_tid=fe6279e0-2f5e-11e7-b-a7-00000aacb35f&acdnat=1493747250_5bbc6814f3c37064cdafcec1a3d2d4fd
[http://dx.doi.org/10.1016/j.toxlet.2016.10.011]

[34]   Academies N. Acute Exposure Guideline Levels for Selected Airborne Chemicals Volume 3 Subcommittee on Acute Exposure Guideline Levels , [cited 2017 May 2]; Available from: https://www.epa.gov/sites/production/files/2014-11/documents/tsd21_1.pdf

[35]   Seiler RL, Wiemels JL. Occurrence of 210-Polonium and Biological Effects of Low-Level Exposure: The Need for Research. Environ Health Perspect [Internet] , 2012 Apr 26; [cited 2016 Dec 22];120(9): 1230-7. Available from: http://ehp.niehs.nih.gov/1104607

[36]   Radioisotopes in Medicine | Nuclear Medicine - World Nuclear Association [Internet] , [cited 2016 Dec 22]; Available from: http://www.world-nuclear.org/information-library/ non-power-nucle-r-applications/radioisotopes-research/radioisotopes-in-medicine.aspx

[37]   Polonium-210 Fact Sheet by Health Physics Society [Internet]. 2010 [cited 2016 Dec 22]. Available from: https://hps.org/documents/po210factsheet.pdf

[38]   Guidance for Industry Estimating the Maximum Safe Starting Dose in Initial Clinical Trials for Therapeutics in Adult Healthy Volunteers Pharmacology and Toxicology Guidance for Industry Estimating the Maximum Safe Starting Dose in Initial Clinical Trials , 2005 [cited 2017 Nov 25];301-

827. Available from: http://www.fda.gov/cder/guidance/index.htm

[39] Water MSDS [Internet] , [cited 2017 Dec 9]; Available from: http://www.sciencelab.com/msds.php?msdsId=9927321

[40] Material Safety Data Sheet Methanol [Internet]. [cited 2017 Dec 9]. Available from: https://wcam.engr.wisc.edu/Public/Safety/MSDS/Methanol .pdf

[41] Material Safety Data Sheet Ethyl Alcohol [Internet]. [cited 2017 Dec 9]. Available from: http://www.sciencelab.com/msds.php?msdsId=9923955

[42] Material Safety Data Sheet Invert Syrup [Internet]. [cited 2017 Dec 9]. Available from: http://www.sugaraustralia.com.au/Documents/MSDS-InvertGoldenSyrupTreacle.pdf

[43] Material Safety Data Sheet Sodium Chloride [Internet]. [cited 2017 Dec 9]. Available from: http://www.sciencelab.com/msds.php?msdsId=9927593

[44] Material Safety Data Sheet Morphine Sulfate [Internet]. [cited 2017 Dec 10]. Available from: http://www.sciencelab.com/msds.php?msdsId=9926150

[45] Material Safety Data Sheet Ibuprofen [Internet]. [cited 2017 Dec 9]. Available from: http://webs.anokaramsey.edu/chemistry/MSDS/Ibuprofen.pdf

[46] Coumarin CAS 91-64-5 | 822316 [Internet]. [cited 2017 Dec 9]. Available from: http://www.emdmillipore.com/US/en/product/Coumarin,MDA_CHEM-822316?ReferrerURL=https%3A%2F%2Fwww.google.com%2F&bd=1

[47] A Study of Median Lethal Dose(LD_(50)) of Doxorubicin on Mice with Intraperitoneal Injection(i.p.)--Journal of Wenshan Teachers' College 2009-04 [Internet]. [cited 2017 Dec 10]. Available from: http://en.cnki.com.cn/Article_en/CJFDTOTAL-WSSZ200904030.htm

[48] Material Safety Data Sheet Aspirin [Internet]. [cited 2017 Sep 12]. Available from: http://bestcareambulance.org/images/MT_aspirin_msds.pdf

[49] Material Safety Data Sheet Caffeine [Internet]. [cited 2017 Dec 9]. Available from: https://www.ingredientstodiefor.com/files/MSDS_Caffeine.pdf

[50] Material Safety Data Sheet Paraquat [Internet]. [cited 2017 Oct 12]. Available from: http://herbiguide.com.au/MSDS/MPAR20_58734-1204.PDF

[51] Material Safety Data Sheet Arsenic Trioxide [Internet]. [cited 2017 Dec 9]. Available from: http://www. sciencelab.com/msds.php?msdsId=9927087

[52] Fentanyl Drugbank [Internet]. [cited 2017 Dec 9]. Available from: https://www.drugbank.ca/drugs/DB00813

[53] Brodifacoum IPCS Inchem Data sheet on Pesticides No. 57 [Internet]. [cited 2017 Dec 1]. https://web.archive.org/web/20131213084637/http://www.inchem.org/documents/pds/pds/pest57_e.htm

[54] ToxTutor - Welcome [Internet]. [cited 2017 Dec 6]. Available from: https://toxtutor.nlm.nih.gov/index.html

[55] Wilson D. Mechanisms of acute toxicity [Internet]. [cited 2017 Dec 1]. Available from: https://ntp.niehs.nih.gov/iccvam/meetings/at-wksp-2015/session4/1-wilson-508.pdf

[56] Relative Toxicity [Internet]. [cited 2017 Nov 27]. Available from: http://venomsupplies.com/toxicity/

[57] Fentanyl: What Is a Lethal Dose? [Internet]. [cited 2017 Nov 29]. Available from: https://www.oxfordtreatment.com/fentanyl/lethal-dose/

[58] Bailey KC. The Biological and Toxin Weapons Threat to the United States , 2001 [cited 2017 May 5]; Available from: http://www.nipp.org/wp-content/uploads/2014/11/Toxin-Weapons21.pdf

[59] Broad A, Sutherland S, Coulter A. The lethality in mice of dangerous Australian and other snake venom. Toxicon [Internet] Pergamon , 1979 Jan 1; [cited 2017 Nov 28];17(6): 661-4. Available from:

http://www.sciencedirect.com/science/article/pii/0041010179902459?via%3Dihub
[http://dx.doi.org/10.1016/0041-0101(79)90245-9]

[60] USACEHR Technical Report 0802. Derivation of Human Lethal Doses. [Internet]. [cited 2017 May 5]. Available from: dtic.mil/cgi-bin/GetTRDoc?AD=ADP494706

[61] Representative LD-50 Values [Internet]. [cited 2017 Nov 20]. Available from: http://biology.unm.edu/toolson/biotox/representative_LD50_values.pdf

[62] Toxins and Known LD50 Values [Internet]. [cited 2017 Dec 6]. Available from: http://www.uab.cat/doc/DL50_biotoxines

CHAPTER 2

Classical Poisons: Arsenic, Hemlock and the Asp

Abstract: Poisons, by their nature, conjure up unpleasant thoughts. To the non-scientist, poisons are associated with cowardly acts, sinister motives, painful death, murder, suicide, and deadly plants and animals. To the scientist, poisons are an enigma – how did they come to be – and they are interesting because the study of their mechanism of action can inform us about how the body works and they are a source of medicinals. Poisons also have been the source of historical intrigues, individual and state-sponsored atrocities, and woven into legends and classical stores. In this chapter, three poisons are considered for their horrible and potent effects, as well as their classical and historical importance. Arsenic can be called the "king of poisons" because of its wide use in the 1800s; it was cheap, easily available, did not have a detectable taste when added to food and drink, it could be given surreptitiously in small doses over time with a cumulative effect, its effects mimicked other common illnesses, and until the Marsh test, it was virtually undetectable in the victim. Hemlock is classically known as the poison that killed Socrates and it is biosynthesized by a plant with familiar non-poisonous relatives including the carrot and parsnip. It causes a slow ascending paralysis starting in the feet and ending in death, and was once designated as the "State" poison because of its use as the means of execution in ancient Greece. The asp is a poisonous viper forever associated with Cleopatra. In fact, we do not know what snake caused her death but the story is classical.

"It is unquestionable that certain words and ceremonies will effectively destroy a flock of sheep, if administered with a sufficient portion of arsenic."[1]

Keywords: Albertus Magnus, Arsenic, Asp, Borgias, Charles Lafarge, Cleopatra, George Bodle, Hemlock, Inheritor's powder, King of poisons, Magic Bullet Theory, Marie Lafarge, Marsh Test, Mees' Lines, Napoleon Bonaparte, Paul Ehrlich, Salvarsan, Socrates, State Poison.

INTRODUCTION

As a means of assessment, poisons can be categorized based on historical use, secretive nature, as an agent for murder, as an agent for suicide, accidental poisonings, and relative to the number of people who have died. Let us see how this affects our thinking about poisonous substances. Surely these considerations won't disqualify our original conclusion, but they will give us further perspective on poisons. I shall describe three poisons to provide some historical perspective and as an introduction to the subject.

The history of the use of poisonous substances is vast and, perhaps, quite subjective and not very scientific. Thus, my perspective for this book includes what I have called the "art" as well as the science of poisons. I vote for arsenic, hemlock, and cyanide as poisons ranked high for reasons other than toxicity based on smallness of lethal dose.

TALES OF THREE POISONS

Arsenic Poisoning

Arsenic could be given the title "King of Poisons", primarily because of its prominent use in Europe during the Roman Empire [1]. It is generally accepted that although impure arsenic was available as early as the fourth century B.C., credit for its isolation (in 1250) is given to Albertus Magnus (1193/1206-1280) [2]. Albertus Magnus (also known as Saint Albert the Great and Albert of Cologne) was a Dominican friar famous for wide knowledge and, significantly to me, for the principle that the study of science was compatible with his religious faith [2] and he established the study of nature as a science within the Christian tradition [3]. He is considered to be the greatest German philosopher and theologian of the Middle Ages and was known as "Doctor Universalis". He wrote prolifically about science, logic, theology, botany, geography, astronomy, mineralogy, chemistry, zoology, physiology, and phrenology. It is said that he was the most widely-read author of his time [2].

The word "arsenic" has an interesting history. Arsenic is said to be derived from the Persian *zarnikh* and *zarniqua* from Syriac and later translated into Greek as *arsenikon* which means masculine or potent and probably referred to what is known as yellow arsenic. *Arsenikon* became *arsenicum* in Latin and then in old French, *arsenic* [4].

Hippocrates, and Dioscorides, respectively, are said to have used and recommended arsenic sulphides in ointments to treat ulcers and abscesses, and as a depilatory [1]. Dioscorides, a Greek physician in Emperor Nero's court in the first century, is credited with describing arsenic as a poison with sinister and perhaps ideal properties that included lack of color, odor or taste when added to wine and other food [1]. Arsenic, given in a large dose, produced death fairly quickly or when given in a sequence of stealthy lower doses it produced an insidious, chronic illness with loss of strength, confusion and paralysis. It was readily available to all classes in society. It could kill quickly by a large dose that mimicked common food upsets, but it was more violent with cramping, vomiting and diarrhea resulting in shock and death [1].

The arsenic of choice was arsenic trioxide and a deadly dose was no larger than pea-sized [1]. It is said that the use of arsenic for politically-inspired murders by Romans was so extensive that in 82 B.C. the dictator Lusius Cornelius Sulla Felix (c. 138 BC-78 BC) (Fig. 1) issued the Lex Cornelia which probably deserves the status as 'the first law specifically against poisoning' [1]. Sculla's lex had provisions against poisoning and those who made, sold, bought, possessed, or gave poison for the purpose of poisoning. The punishment specified by the law, according to one authority was confinement for life; the convicted person was considered to be civilly dead [5].

During the Middle Ages in Italy, the most infamous of poisoners were the Borgias - Rodrigo Borgia a noted Renaissance person who became Pope Alexander VI, and his son Cesare. Many historians include Cesare's half- sister, Lucretia; some say she was innocent but her name is now linked irrevocably with poisoning by

Fig. (1). Lusius Cornelius Sulla Felix, Roman dictator credited with perhaps the first law criminalizing poisons and poisonings.

arsenic [1]. Lucretia Borgia (1480-1519) deserves better than her popular image today which is one of alleged involvement in various political scandals, sexual intrigues, and poisonings, which does not tell the whole story of her life [6, 7]. Indeed, the story of corruption and death associated with her and her family has been added to by rumor and exaggeration. Her life after the death of her father and during her third marriage included developing a strong Christian faith which can be regarded as a triumph of redemption.

The long history of arsenic poisoning includes rumors that Napoleon Bonaparte's death in 1821 was due to repeated poisoning with arsenic; this has not been confirmed by analysis of hair samples from his corpse [1]. Claire Booth Luce (1903-1887), a member of Congress and United States Ambassador to Italy (the first woman to be appointed to a major ambassadorship) was the victim of poisoning from a bizarre circumstance. During her years (3+) as Ambassador to Italy she was poisoned and medical analysis eventually determined that she suffered poisoning by arsenate of lead from paint dust falling from the stucco of the bedroom ceiling of her Ambassador's residence in the 17th century Villa Taverna in Rome. The story was featured in Time magazine (June 23, 1956) [8]. She recovered from this illness and died years later from brain cancer which might, or might not, have been related to the arsenic poisoning. Arsenic is a known carcinogen.

Arsenic also appears in records about Chinese medicines dating to 200 BC, including in the first traditional Chinese book of medicine "*Shen Nong Ban Cao Jing*". Arsenic was said to have been a common component of Indian "Ayurvedic" herbal medicines. An elixir made from the "essence of five planets" may have contained arsenic and certainly other metals and was promoted as a medicinal. Arsenic had an early and leading role in the development of targeted medicines with the first effective treatments for the great scourges of syphilis and trypanosomiasis, and arsenic trioxide is currently still approved for treating a type of refractive leukemia. As early as 1931, based on the medicinal Fowler's solution that was developed in 1878, an arsenic compound was used to lower the white blood cell count of patients with chronic myelogenous leukemia [4]. The use by Paul Ehrlich of a chemical compound of arsenic was the first targeted synthesis of a chemical directed at a specific microbe, and was a forerunner to and the only treatment for infectious agents until the discovery and development of penicillin.

In 1918, organic arsenic was incorporated into the chemical warfare agents Lewisite and Adamsite but this was too late for them to be used in WWI; however, they are still potential agents for warfare and bioterrorism. Paracelsus wrote his own pharmacopoeia and noted the therapeutic effects of chemical elements including arsenic. He is said to have been the first to provide specific directions for the preparation of metallic arsenic including a balm from white arsenic that was used by "barber surgeons" of that era for treating wounds and ulcers of various types [4]. During the Renaissance the "art of poisoning" flourished to the extent that poisoners were sought and for payment the deed was done. Famous poisoners included the Borgia family, Pope Alexander IV, his son

Cesare, and Cesare's half sister, Lucretia, and in Italy, Giulia Toffana who concocted cosmetics containing arsenic. Indeed, Toffana and her daughter, Girolama, were executed in 1659 for their involvement in what is said to have been the "death of several hundred men" [1].

A cartoon from the 1800s gives insights into arsenic poisoning in that century (Fig. **2**). The "paterfamilias" is, of course, the father or head of the household, and he is warned about the dangers of arsenic (and possibly other things) in "sweets" that might find their way to the family dinner table. The old meaning of lozenge was a small medicated candy. Was this the forerunner of Halloween candy scares in modern days? The cartoon was drawn because of the practice of adulteration of food, typified by an infamous case in Bradford in 1858 [9, 10]. Many people were made ill and more than 20 died, including children, after consuming adulterated peppermints. One Joseph Neal had been replacing some of the sugar with plaster of Paris to save money. Unfortunately, on one occasion apparently without knowing it, he had used another white substance, arsenic. Before this tragic event, in the summer of 1855, the English Select Committee on the Adulteration of Food, Drink, and Drugs (created by Parliament) had discovered and revealed to the public that candies were being colored using toxic minerals including a vivid green that was made using copper acetoarsenite!

THE GREAT LOZENGE-MAKER.
A Hint to Paterfamilias.

Fig. (2). An old cartoon warning about the dangers of poisoning[2].

It is stated that the lozenge problem began on October 18, 1858 when a man named Joseph Neal went to the shop of a druggist named Charles Hodgson to obtain twelve pounds of a substance called "daff" [11]. Oddly, daff, as it was known by confectioners, was said to have been plaster of Paris, powdered limestone, or sulphate of lime. Tragically, however, in this case a form of arsenic was substituted. This is how the unfortunate substitution unfolded. When Neal arrived to make his purchase, Hodson was ill in bed but had hired a helper Goddard who ultimately was informed the "daff" was in a cask in the corner... a white powder. Consequently, the lozenges were concocted with 40 pounds of real sugar and 12 pounds of arsenic plus 4 pounds of "gum-water". Both Neal and Goddard also became ill, and neither was ever suspected of deliberately poisoning the batch. Some of the lozenges were delivered to a dealer called "Humbug Billy" (I am not making this stuff up) who didn't like the color but when the price was reduced, he accepted them. Soon, victims began to appear. They experienced "great retching, vomiting, pain and burning of the throat, intense thirst, pain in the abdomen and diarrhea" [11]. Cholera was at first suspected, but the common factor, as more cases developed, was obviously the lozenges. It is said that policemen were sent out to give warnings up and down their beats and that "two bellmen [travelled] through all the streets of the town from eleven till five o'clock, warning the inhabitants" [11]. Charges were filed, but dropped against Goddard and Neal and Hodgson were tried and acquitted. At trial, a physician, Dr. John Bell, testified to identify arsenic as the poison, and this was confirmed by a prominent chemist, Felix Rimmington. It was stated [11] that: "Each lozenge it [was] supposed... contain [ed] 9/12 grains of arsenic, and 41/2 grains [were] considered to be a poisonous dose, each lozenge was sufficient to poison two persons... [which meant] there was sufficient poison distributed... by Hardarker ... [to] kill nearly 2,000 persons." In all 200 persons were made ill, as previously stated. This gruesome event was instrumental in some good: the passage of the Pharmacy Act of 1868.

Arsenic May Have Been Used by Nero for Murder

History records that Nero used arsenic to kill his stepbrother, Britannicus, on his path to becoming Emperor [4]. It has been reported that Nero had the corpse covered with gypsum to cover-up the effects of the poisoning on the skin; however it rained while the body was being carried to the Forum and washed off the gypsum to reveal the effects of arsenic poisoning. I found no details of the circumstances of the alleged poisoning.

Arsenic Poisonings in 19th Century England

In an article in 2013 in *The New Yorker,* Joan Acocella wrote that a third of all poisonings in England throughout much of 19th century were from arsenic [12]. The reason given, besides arsenic's effectiveness, was its ready availability and its low cost- a tuppence for a half ounce. A tuppence (two pence) was worth about 4 cents U.S. in that era. This author states that other reasons may have been: the publicity that newspapers gave to murder by arsenic; the fact that the symptoms of diarrhea, vomiting, and abdominal pain were delayed and similar to other ailments; and the poison could be given in repeated, small doses over many days [12]. Poisoning was often done by women (poisoning their husbands) – so frequently that the House of Lords attempted to pass a law forbidding women to purchase arsenic [12].

A sensational case of arsenic poisoning was told in the book *The Inheritor's Powder*, a Tale of Arsenic Murder and the New Forensic Science, by Sandra Hempel [13]. During the first half of the 19th Century, Europe was swept by a veritable epidemic of arsenic poisoning. It was so prevalent that arsenic became known as "the inheritor's powder" [13]. Arsenic was tasteless in food or drink and there was no way to forensically identify arsenic is suspected sources.

The poisoning murder of George Bodle on November 2, 1833 changed all this. Bodle and members of family together with their servants became ill after a breakfast meal. A physician, John Suther was called but three days later, George Bodle died in agony at his farmhouse in Plumstead, England. Poisoning was suspected; the case drew national attention; and several heirs were suspected of poisoning the food served to Bodle. There were bickering heirs, a bumbling policeman who was also said to be a drunkard, and a brilliant chemist who became noted for creating a new analysis procedure that became known as the Marsh test that could accurately detect arsenic in food and drink. These facts are told by Sandra Hempel [13]. Hempel further related that doctors in that era frequently misdiagnosed the signs and symptoms of arsenic poisoning as cholera, malaria, or dysentery. This case became a landmark in medical-legal history. Doctors involved in the Bodle case became suspicious and collected Bodle's vomit. They believed that tainted coffee was the vehicle. Packets of arsenic were found in the possession of Bodle's grandson, John, who was arrested and tried for murder. An autopsy was performed and James Marsh, who was an assistant to the famous Michael Faraday, was called as an expert witness to testify but was unable to convince the jury that John Bodle was guilty of arsenic poisoning. This failure sent Marsh on a successful search for a new test procedure for arsenic. He was so successful that Hempel calls this test "the first major advance in modern chemical toxicology." Marsh developed the arsenic test in 1836, some three years after the Bodle trial [14]. The test procedure involved adding a tissue sample or body fluid to a glass vessel containing zinc and acid. Arsine gas was produced when arsenic was present and when ignited a silver-black deposit was formed on the glass [14].

The test was very sensitive and minute traces of arsenic could be detected. This is relevant to the Marie Lafarge arsenic poisoning case of 1840 that is described below. To complete the story, John Bodle who was George Bodle's grandson and heir was not convicted at the trial. He was freed but later confessed to the murder.

The Intriguing Case of Marie Lafarge

McClure's Magazine in 1912 published an account written by Marie Belloc Lowndes of a trial in 1840 for murder by arsenic poisoning [15]. This one case embodies most of the elements of classical arsenic poisoning. There is the secretive nature of events around the suspicious event, possible repeated poisoning, inability to taste the poison in food and drink, the non-diagnostic nature of the signs and symptoms shown by the victim, the difficulty in detecting arsenic in the body leading to doubt about whether poisoning even occurred, and the ready-availability of arsenic.

The story of this case truly reveals the essence of "the art and science" of one poisoning and provides the means of addressing the chemical and biological effects of a poison, the insidious nature of the poisoning event, the complex characters of victims and suspected perpetrators, the difficulty of forensics in identifying poisons, and the effects of sensationalism and popular opinion on the outcome of trials. Additionally, this case affords the opportunity of addressing a very potent toxic chemical, arsenic; provides historical perspective into forensic toxicology of the mid-1800s; and colorfully portrays the influence of the courtroom and public opinion on trials which continues with high-profile trials of today.

The Overview

The sensationalized Lafarge trial in France in 1840 was followed extensively on a daily basis by the public through newspaper coverage. The account provided here is my interpretation and is taken primarily from the story by Marie Belloc Lowndes [15] plus three other sources written by Geri Walton [16], Brandy Schillace [14], and William Jensen [17]. A young, beautiful, gifted, French woman named Marie Lafarge was charged with murder by arsenic poisoning. The victim was Marie's husband, Charles Lafarge who was often described as "coarse and repulsive". This case exemplifies characteristics of poisonings in that era, and the complexities presented to the courts, expert witnesses, juries, and the public. With different specifics, this case typifies in extreme ways, the ideas of this book that poisonings involve both art (in its broadest sense) and science. Was Marie a victim of circumstantial evidence or a clever murderer? We shall never know. As happens today, trials can (unfortunately) be influenced by popular opinion and by sensationalism during the trial. In this case, the lack of a sure way to detect arsenic and circumstantial evidence contributed to the conviction of Marie.

The Setting

On August 6, 1839 Marie and Charles Lafarge, married only for four days, left Paris for *Les Glandiers*, the French country estate house of Lafarge (Fig. **3**), as interpreted by Marie Belloc Lowndes [15]. Soon after their arrival, Marie locked herself in a room and a scene erupted involving Marie, Charles, and Lafarge's mother. This is important because it helps to understand these characters including the mind-set of Lafarge's mother, Madame Lafarge. She disliked Marie and "…though she is already violently prejudiced against her Parisian daughter-in-law, [she] is full of that mingled shred sense and feeling of self-respect that is one of the foundations of the French character" [15]. Charles is portrayed as unreasonable, and coarse… "Marie, open the door! What do you mean by locking yourself in, in this way? Do you not understand that I am your husband, and that I have the right to order you to open the door?" [15]. A letter written by Marie was pushed by Marie from inside the locked door and taken up by Charles who became enraged as he read it. The letter was used later at trial to help establish the

Fig. (3). The estate of Charles Pouch-Lafarge in Glandier, a hamlet in France (public domain, courtesy of Bibliotheque nationale de France).

circumstantial guilt of Marie. The letter included… "I love another man!... I am deeply ashamed of my wickedness!... I only ask you to let me go away" [15].

Geri Walton [16] gives an account of prior difficulties in this brief marriage: "...before they reached Glandier, Marie became disgusted with her husband's brutality. She later described herself as 'utterly ignorant of wifely duties and marital relations,' an assertion that one newspaper claimed was 'hardly credible as applied to a French woman then 23 years of age'..." At age twenty-three Marie remained unmarried, and to make matters worse, there were no suitors on the horizon. Her uncle had grown tired of supporting her... and became focused on finding her a husband...[He] hired a marriage broker...The broker was the same broker... that was hired by Lafarge to find him a wife." [16]. Supposedly, because of his brutality, a painful scene unfolded between the newlyweds. She refused to share her husband's bed at the hotel in Orleans, and as told by Walton, 'she [Marie] became hysterical, and locked herself within her own apartment' [16]. The newlyweds worked this out and reconciled. However, it was short-lived. Walton provides the following perspective: "Marie saw that both she and her relatives had been grossly deceived as to the fortune of her husband. Rather than some luxurious estate, she found a rat-infested, crumbling mansion, and instead of wealth she found her husband facing considerable debt" [16].

The Principal Characters

Marie-Fortunee Capelle (Marie Lafarge) (Fig. **4**) was charged with the murder. She was born in Picardy, France in 1816. As told by Geri Walton [16]: "Her father was an artillery officer, and her grandmother was rumored to be the illicit love child of Stephanie Felicite, better known as Madame de Genlis, and Louis Philippe II, Duke of Orleans. If that was true it made Marie a descendant of Louis XIII of France... Marie was described as 'not greatly blessed with beauty'. She found herself at the age of eighteen the adopted daughter of her maternal aunt... Marie was sent to the best schools... However, she was always aware that she was nothing more than a poor relative... Marie dreamed of marrying some rich well-to-do aristocrat, but as she had no say in who she would marry and because her dowry- 80,000 francs- while considerable, was not that impressive based on her family's status." [The internal quotation marks are in the original]. A somewhat different perspective of Marie Lafarge is inferred by Marie Belloc Lowndes [15]: "Charles Lafarge fell in love at first sight with the elegant, beautiful, and accomplished Parisian girl... Still Marie never accused him [Charles] of active unkindness; all she said was that during her honeymoon her eyes were opened to the fact that she had married a man very unlike herself, one who cared nothing for books, for music, for society, or for anything except business. The high-strung, romantic girl also found that Les Glandiers was entirely different from what she... had been led to expect... Marie liked everything done in a simple yet elegant manner."

Fig. (4). Marie Lafarge (public domain, courtesy of Bibliotheque nationale de France).

Charles Pouch-Lafarge, who was allegedly poisoned with arsenic, according to Geri Walton [16]: "… was a coarse and repulsive 28-year-old man… not having much luck … he had married and his wife had died shortly thereafter… his father had purchased property in the hamlet of Le Glandier… it had fallen into disrepair… Lafarge turned part of it into a foundry… [resulting in him] falling into massive debt and being on the verge of bankruptcy… Lafarge decided to find a wife that could help him financially… It [the marriage of Charles and Marie] was a loveless marriage entered into from sordid motives on both sides." Marie Belloc Lowndes [15] describes Charles as follows: "A good-looking man…Lafarge's first impulse, on reading the strange and rather crazy epistle, is to batter down the bedroom door and kill the woman [Marie] who is behind it... he is full of rage, and beside himself with jealously… Lafarge was twenty-eight, member of an honorable provincial family, and an iron-master in a fair way of business… he owned a delightful country house – castle, in fact, named Les Glandiers – in that beautiful district… Marie was soon won over to the view that Charles Lafarge concealed a heart of gold beneath his rather rough exterior, and that his love would soon bring forth hers… At no time had he struck those about him as particularly refined or well-bred; but now, when he was sure of his beautiful young wife, he showed himself in his real colors, that is, as a rough and rather boorish individual, determined to have his way in everything. Still Marie never accused him of active unkindness; all she said was that during her honeymoon her eyes were opened to the fact that she had married a man very unlike herself."

The Intrigue

Geri Walton [16] described significant events and provided five references which she used as the bases for facts about significant events and from which she wrote some direct quotes. I provide some quotes from her that included her quotes as indicated by sub-quotation marks. Marie wrote a letter of recommendation for Chares which he took to Paris in December, 1839 in a futile attempt to find investors for a claimed new process for smelting iron. Marie had her portrait drawn and sent it along with five small cakes (which she had baked) in a parcel that witnesses later testified they had seen her seal. However, when Charles received the parcel it was unsealed and there was one large cake rather than five small ones. Walton writes (the interior quotes are hers): "…when he received the parcel he was delighted, 'partook of [the] cake, and soon after he became excessively ill, and was compelled to return home'… where he arrived on 5 January 1840… Lafarge suffered from for more than a week and died on the 14th of January leaving everything to his poor widow… In the middle of January 1840, just as Marie was about to inherit all of Lafarge's wealth, she was arrested by police on an astonishing charge: Police claimed she had poisoned her husband. Moreover, her method of murder was supposedly 'by administering arsenic to him during his illness.' With a charge of murder levied … a search of Marie's house was conducted, which in turn resulted in more charges against her" [16].

The Alleged Arsenic Poisoning and the Trial

The trial was delayed a bit and occurred on July 9th, 1840. On a charge of theft (not addressed because it was not directly related to the alleged poisoning) Marie was quickly found guilty and she was then arraigned on the charge of murdering her husband. Geri Walton [16] provides the following quotation (with the sources imprecise as one of four references listed): "These proceedings were 'unparalleled in the records of jurisprudence,' because for the first time, forensic toxicological evidence would be crucial in getting a conviction." This is an important reason why I document this matter extensively in this book about the art and science of poisons.

Walton states [16]: "The public Prosecutor wanted to 'convict the accused', and the Judge, using copies of previous examinations, also attempted to trap the widow Lafarge into giving a contradictory statement. In addition, evidence from the prosecution showed Lafarge's illness began after he ate the cake and that one large cake arrived rather than five small ones. Moreover, the prosecution alleged it was the widow Lafarge who packed and sent the 'medicated cake', yet she was not the only one suspected of murdering Lafarge." There is a complication created by one Denis Barbier, whom the defense alleged was the perpetrator of the poisoning and that he had the opportunity to switch the five (innocent cakes baked by Marie) for one poisoned one.

An important issue for our telling of the story is proof that Marie had access to arsenic. It was determined that she purchased arsenic in December from a druggist with a written statement that "she required it [arsenic] for the purpose of destroying rats" [16]. She made a second purchase from the same druggist on January 5th (with a written statement similar to the first) and this was after the ill Charles had returned home from Paris. Barbier claimed that Marie had asked him to buy arsenic on more than one occasion and had "begged him to say nothing about it" [16]. Trial testimony indicated that Madam Lafarge (Charles' mother), upon learning this while Charles was ill, but yet alive, became suspicious and Marie was watched. Walton writes that: "One evening Lafarge asked for chicken broth. His sister made him some and left it on a mantelshelf. A friend of the family, a mademoiselle Le Brun, later testified that she saw Marie 'reach out her hand toward the bowl and put a white powder into it, stirring the fluid with her finger." It was afterwards shown that a sediment found in the bowl contained arsenic" [16]. Arsenic in large quantities was found throughout the mansion, and Le Bruin stated he saw Marie "…take a glass of wine, take something out of a drawer, and mix it into the wine with a spoon… when the powder in the drawer was checked, a chemist declared it to be arsenic" [16]. After his death, an examination of Lafarge's body was immediately conducted with the following results reported: "… the viscera were removed, placed in unsealed vases, and sent to the chemists at Brive. No precaution was taken to prevent those organs from being tampered with. The chemists of Brive declared that they found arsenic in the stomach… However, another analysis was completed by Limoges chemists. Their results proved opposite, and they declared positively that the most minute tests, including Marsh's process, failed to disclose the slightest trace of arsenic" [16]. During the trial, based on these conflicting results, the court ordered Lafarge's body exhumed and a French toxicologist (Mathieu Joseph Bonaventure Orfila) was enlisted and declared "he found arsenic in Lafarge's stomach and its contents, but not in his tissues" [16]. This could be interpreted as evidence that someone had sprinkled arsenic on the viscera.

Marie Lafarge's trial lasted sixteen days and a verdict of guilty with extenuating circumstances was brought in by the jury. She was sentenced to "imprisonment for life with hard labour and exposure to the pillory…because of trial irregularities, an appeal was lodged in her behalf, but it 'was rejected… Lafarge was then imprisoned and remained imprisoned for twelve years, until her health gave way, and, in consideration of her debility, she was liberated in 1852'… [but]

only survived a few months'... Was she guilty?" was the summation by Walton [16]. Indeed, Walton [16] assessed, in part, that an examination of the trial and its circumstances in 1842 concluded: "that Barbier did not have the best of character, was said to have 'lived by forgery, and was the accomplice of Lafarge in some very shady transactions, by which that unhappy man sought to cover his insolvency. Barbier had [also] conceived a violent hatred against Madame Lafarge, as her presence was likely to hinder his nefarious practices, and especially to weaken his hold over his companion in crime'...Furthermore, it was noted that Barbier had unrestricted access to the mansion, the chicken broth, the wine glass, the drawer, and even Lafarge's corpse and stomach, which would have allowed Barbier to introduce arsenic into any of these places."

The account of the Lafarge matter by Lowndes [15] is considerably more poetic and detailed than that cited above. Regarding the issue of the possibly poisoned cake(s), Lowndes states: "Marie further proposed that Charles should eat one of these cakes at twelve o'clock on a certain night, she at the same time eating a similar cake. She also added, in the same letter, that he had better not tell any one of this sentimental refection. Lafarge received the letter, and the box [containing the cakes]. But he did not open it himself; instead, he told one of the hotel servants to do so… there was only one large cake in the box. Lafarge then broke off a small piece of the cake, and… ate it. That night he was taken violently ill [proven by the hotel records]. The rest of the cake was thrown into a drawer and… one of the hotel servants ate a piece of it, and was also taken ill in exactly the same way as Lafarge… Marie Lafarge [based only on testimony of her mother-in-law who reportedly hated her]…began to show the most surprising uneasiness as to her husband's health… Lafarge was far from well when he arrived at Les Glandiers… he was so unwell that he had to go to bed at once. Marie herself brought him his supper, which consisted of cold truffled fowl, and she and he ate it together… No sooner had he eaten than Charles was taken terribly ill, with symptoms … believed later to have been those of poisoning, though at the time no one suspected such a thing. These symptoms were a terrible heat in his throat, horrible gnawing pains, and an awful sense of icy coldness… long, weary days wore themselves away. Lafarge lay ill, eating very little… more or less prepared, and always handed to him, by the devoted Marie." A woman (Madame Brun) who was a friend of Madame Lafarge entered the household at this time and may have been a kind of spy for her. She stated that: "…she watched the unsuspecting Marie very closely, and she noticed – or declared afterward that she noticed – that whenever young Madame Lafarge was about to hand her husband anything to drink she always put into the cup a spoonful of white powder. Lafarge daily grew worse and worse, and at last his wife became so much alarmed that she implored the local doctor … to call in a specialist… It cannot be stated too clearly that up to that time no one, least of all the doctor attending

Lafarge, suspected poison. The doctor … made light of Marie's fears, and told her that Charles Lafarge had been subject to these attacks from childhood."

The arsenic at Marie's orders had been made into a paste and placed around to kill the rats that infested the mansion, but without effect [15]. Another doctor was called in to attend the suffering Charles. He apparently believed the story that Madame Lafarge told him that "Marie was slowly but surely poisoning her unhappy son!" … He decided on what most people will agree was a very cruel course: he decided, that is, that it was his duty to tell the now dying man that he was being done to death, and by the wife in whom he implicitly trusted…. Lafarge took the news calmly, and told him, as confirmatory evidence, of the illness he had had in Paris after eating the cake that Marie had sent him!… [he] still spoke of his wife in terms of adoring affection, he still would not allow her out of his sight, and he never took any food, excepting from her hand. Against this weight of evidence there is one contradictory statement made by Lafarge's sister, but we shall see how slight and fanciful that statement is… at last there came the dread moment when the dying man's wife, his mother, and his sister, together with various other relation, came and knelt around his bed. Suddenly Lafarge muttered to his sobbing mother: 'You hurt me – please go away!'… Lafarge hoarsely exclaimed 'I want to drink!' Marie rushed forward with some water. But Lafarge opened his eyes, and – still only according to his sister's evidence – a dreadful smile came over his face as he pushed the glass his wife's hand held out to him from his lips. It is a rather strange fact that Marie left the room before her husband's death."

There is much more in the McClure's Magazine story by Marie Belloc Lowdes [15] from which I have extensively quoted as the only means of accurately providing the issues surrounding this nearly two-hundred-year-old tale of arsenic poisoning (if indeed it was arsenic that killed Charles Lafarge). The circumstances of the alleged administering of the arsenic unnoticed in food and drink, the repeated poisonings, the ready-availability of the arsenic, the symptoms of the victim, the difficulty of the doctor's in following those symptoms to a suspicion of poisoning, the possibility, not mentioned by other authors, that Lafarge may have been delusional from the effects of arsenic – all illustrate what I wish to portray about the insidious nature of this poison, arsenic.

The Lafarge case also helps us understand human nature: that of the victims, the criminals, those people with various interests who are near the victim in relationship and in circumstances, attorneys, expert forensic witnesses, police and judges and juries in murder cases, and even the media and the public at large. Truly, there is more of interest to be found out when considering poisonous substances, than their chemistry and means of affecting the human body.

Arsenic, the Poison, and its Toxic Mechanisms

Arsenic is a naturally occurring element that exists in both organic and inorganic forms. It is of concern as an environmental pollutant and can contaminate drinking water. Historically, it is said to be the "favorite and most famous of all poisons" [17]. As a poison it is usually in the form of its oxide As_2O_3 which is sometimes called white arsenic. World Health Organization guidelines for arsenic in drinking water are 10 µg/l as the safe limit and 50 µg/l as the maximum permissible limit [18]. Chronic exposure to low levels of arsenic (with arsenite, As^{3+} being most toxic) is "associated" with many disorders including cancer, diabetes, hepatotoxicity, neurotoxicity, and cardiac dysfunction [19].

The toxicity of arsenic is complex. It is described as a metalloid (metal-like) substance widely present in soil and water in low amounts. It has inorganic and organic forms and several oxidation states. It is a carcinogen with chronic and acute toxicities. It is metabolized in the body including by methylation and changes in oxidation state which affect toxicity in complex ways.

The toxicity of arsenic depends on its chemical state and a publication by Michael Hughes is a good source for LD_{50} values [20]. For example, arsenic trioxide (a trivalent arsenical) by the oral route, experimentally in the mouse ranges from 15 to 48 mg As per kilogram of mouse body weight (as reported in three publications) cited by Hughes. However, in the hamster, monomethylarsonic acid *via* the intraperitoneal route in the hamster has a LD_{50} of only 2 (in the same units). It is generally agreed that acute toxicity of trivalent arsenic is greater than for pentavalent arsenic. The human lethal dose range is generally accepted to be 1 to 3 mg of Arsenic per kg of body weight which equates to a total dose of approximately 70 to 210 mg [21]. Methylated, trivalent arsenic has greater acute toxicity and this is accepted as evidence that methylation of arsenic, which can occur in the body, is not (solely) a detoxification mechanism, as has been proposed.

Chronic exposure to comparatively low levels of arsenic produces toxic effects throughout organ systems in the body (Fig. **5**). This exposure commonly occurs *via* drinking water in regions where arsenic is a significant water contaminant. However, in the context of poisoning, arsenic has been noted to be chronically administered in a sinister and surreptitious manner. Skin lesions are commonly listed and are characterized by hyperkeratosis and hyper-pigmentation.

To understand the cellular mechanisms of arsenic toxicity distinctions must be made between the trivalent and pentavalent forms. Trivalent arsenicals react with sulfur-containing molecules that are important reducing agents in cells. This includes reduced glutathione (GHS). It also reacts with an enzyme cofactor lipoic acid that is necessary for certain enzymes, including one called pyruvate dehydrogenase, to function. When this enzyme is blocked, an important source of the intermediate that feeds the principle mechanism by which ATP is produced in cells is blocked. These mechanisms are well established [20]. Pentavalent arsenic can replace the phosphate group that is normally the constituent on many intermediates in the cellular metabolism of carbohydrates. These arsenate-substituted compounds are so similar to the real substrates that they are accepted by the enzymes which then, however, are poisoned (blocked from their natural activity). Arsenate can also replace phosphate in an essential anion exchange mechanism in red blood cells and poisons this function. Arsenate also "uncouples" the formation of ATP by a process called "arsenolysis". By substituting itself (as arsenate for phosphate), a subtle change is made in the structure of essential metabolites in a chain of events required for glycolysis (a process that makes ATP without requiring oxygen). ATP is depleted in laboratory experiments; however, arsenic is probably not very effective in depleting ATP in human red blood cells. Understanding of the exact mechanisms of arsenic is complex and not completely understood and the mechanisms of toxicity of pentavalent arsenic are not settled science [20].

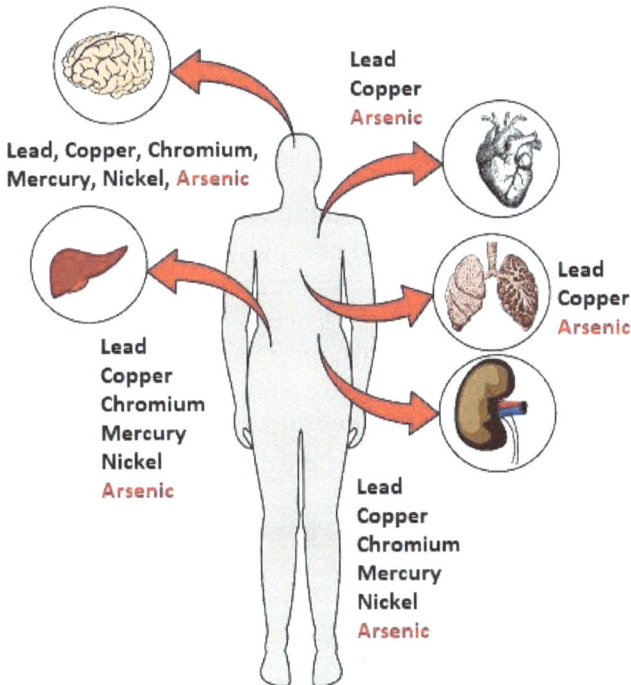

Fig. (5). Sites of arsenic toxicity in the human body. Arsenic has sites in common with classical heavy metals.

Mees' lines (also called leukonychia striata) can appear in finger and toe nails after arsenic poisoning (Fig. **6**). They usually are bands of white across nails that move forward as the nail grows. They can present with poisoning by some other metals, and sometimes result from renal failure and, thus, they are not truly specific for arsenic. Arsenic can be detected in urine and blood samples and is useful for detecting environmental or occupational exposure which is usually comparatively low level. Organic arsenic compounds are eliminated rather quickly in the urine. Inorganic forms of arsenic are usually converted to organic compounds and excreted also.

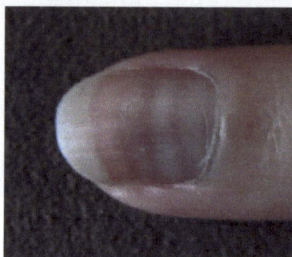

Fig. (6). Mees' lines can be a sign of chronic arsenic poisoning, but are not specific for arsenic.

Tests are available for quantifying arsenic in blood, urine, hair, and nails. The urine test is more reliable than the blood test except when exposure is quite large. Arsenic levels over the past half-year to one-year can be measured by tests of hair and fingernails.

In the following assessment, primary attention will focus on acute effects of high levels of exposure and not on effects of chronic, low level exposure. Acute arsenic toxicity starts with nausea, vomiting, abdominal pain and severe diarrhea but clinical manifestations can occur in virtually any organ. There can be acute psychosis, diffuse skin rash, toxic cardiomyopathy, and seizures. Hematological abnormalities can occur and renal failure and pulmonary edema are said to be common. Neurological manifestations include peripheral neuropathy or encephalopathy [22]. These symptoms help explain the difficulty, especially in the 18th century, that physicians had in diagnosing arsenic poisoning. A recommended source for toxic information is one available from the Agency for Toxic Substances and Disease Registry (ATSDR) [23]. Arsenic targets many enzymes that are found in nearly all organ systems. Arsenic is "strongly associated" with lung and skin cancers. Arsine gas can cause hemolytic syndrome. Gastrointestinal effects occur primarily after ingestion of arsenic. Acute arsenic toxicity can cause hepatic necrosis and elevated liver enzyme levels. Two specific mechanisms of arsenic toxicity directly impair cellular respiration. Arsenic binds to sulfhydryl (SH) groups; thus, it inactivates SH-containing enzymes. Arsenic with 3+ valence is particularly potent for this. Disruption of SH-containing enzymes inhibits the oxidation pathways for pyruvate and succinate in the tricarboxylic acid (TCA) cycle. This impairs gluconeogenesis (synthesis of glucose). Arsenic 5+ valence can substitute for phosphorous in many biochemical reactions in cells. Replacing phosphorus for arsenic leads to rapid hydrolysis of adenosine triphosphate (ATP) which is the universal source of energy in cells. Arsine gas acts differently. Upon inhalation, arsine is fixed to red blood cells and produces irreversible membrane damage leading to hemolysis of red blood cells with dire consequences. At low concentrations, arsine causes a dose-dependent intravascular hemolysis; at high concentrations it produces multisystem cytotoxicity.

The GI tract effects of arsenic are generally only seen with ingestion. The basic lesion in the GI tract is increased permeability of the small blood vessels which causes fluid loss and hypotension. Extensive inflammation and mucosal and submucosal necrosis of the stomach and intestine can be so extensive as to cause gut perforation. Hemorrhagic gastroenteritis can develop with bloody diarrhea.

Severe acute arsenic poisoning can be accompanied by acute tubular necrosis causing acute renal failure from hypotensive shock, hemoglobinuric or myoglobinuric tubular injury, or direct effects on tubule cells and glomerular damage can cause proteinuria. The kidney, however, is not considered to be a major injury target for chronic arsenic toxicity.

Injury to the cardiovascular system varies with many factors including the magnitude of the arsenic dose. The fundamental lesion in acute toxicity from large amounts of arsenic is diffuse capillary leakage leading to generalized vasodilatation, transudation of plasma, hypotension, and shock.

Neurological effects of acute severe poisoning include ascending weakness and paralysis. Cranial nerves are rarely affected even when poisoning is severe. Encephalopathy has been described for acute and chronic exposures. Onset of neurological symptoms can initiate within one to three days after acute poisoning and develops more slowly with chronic exposure. Neuropathy is primarily due to destruction of axonal cylinders (axonopathy).

Both acute and chronic arsenic poisoning can damage the hematopoietic system and bone marrow depression with pancytopenia may occur. Anemia and leucopenia commonly occur and leukemia can be normocytic or macrocytic and basophilic stippling can be present in smears of peripheral blood.

Arsenic formerly was a constituent of some cosmetics and has been used to protect crops from various pests. A form of arsenic known as copper acetoarsenite was used as a pigment called Paris green. In past centuries when coal fires and gas lighting was common, hydrogen gas could be released and combine with the arsenic in the Paris green that was sometimes used in wallpaper to create arsine, a toxic gas. It is said that the mold *Scopulariopsis breviculis*, which sometimes grows in damp wallpaper, can produce arsine from the arsenic in Paris green if present in the wallpaper [22].

Inorganic arsenic is classified (based on epidemiological and not on animal studies) by the International Agency for Research on Cancer (IARC) and the U.S. Environmental Protection Agency (EPA) as a known human carcinogen (its highest classification for carcinogenicity). The mechanism of carcinogenicity is not known [20].

Arsenic in widely distributed in nature but its abundance in soil and water is low. For example, its abundance in the Earth's crust is estimated to be 0.0001% where it is mostly associated with ores of metals like copper and gold. Arsenic and several methylated (organic) forms arsenate are found naturally in waters of streams and lakes and in some subsurface waters.

It is noteworthy that microorganisms have been found that use arsenic for energy generation by oxidizing arsenite or reducing it. Arsenite cycling may occur in the absence of oxygen and can contribute to organic matter oxidation, and in aquifers, arsenic may be mobilized from the solid to aqueous phase causing contamination of drinking water [24].

Arsenic Forensics

Tests to detect arsenic were not available until the late 1700s and these tests generally relied on forming precipitates as colored compounds. When applied forensically to tissue samples or stomach contents the tests were quite unreliable because interfering substances also formed colored precipitates. An example of the old form of the procedure was the Hahnemann's test which was replaced by the Marsh test (Fig. **7**) [17].

As a microbiologist, I cannot leave the story of arsenic without briefly relating a medical use of arsenic. In 1910, Paul Ehrlich introduced the arsenic-based drug called Salvarsan as a treatment for syphilis. The hypothesis that he used for this discovery was based on which he called his "magic bullet theory." He combined a poison with a substance with an affinity for the desired target in the human body at the cellular level. He chose to use a dye for the latter and after many trials, he was successful with a compound he first called simply number 606. It rapidly became the most widely prescribed drug in the world, and remained as the most effective treatment for this venereal disease until the use of penicillin in the

1940s. The exact structure of Salvarsan was controversial and Ehrlich postulated that it contained a particular feature, arsenic doubly bonded to arsenic, a feature that is quite unstable chemically. In 2005 scientists discovered that the true structure of this agent, the first of its kind, a chemically-synthesized agent designed to target the cause of an infections, was a mixture of cyclic bonded arsenic species that slowly releases an oxidized species that is the actual toxin, with such exquisite specificity for the bacterium that causes syphilis [8].

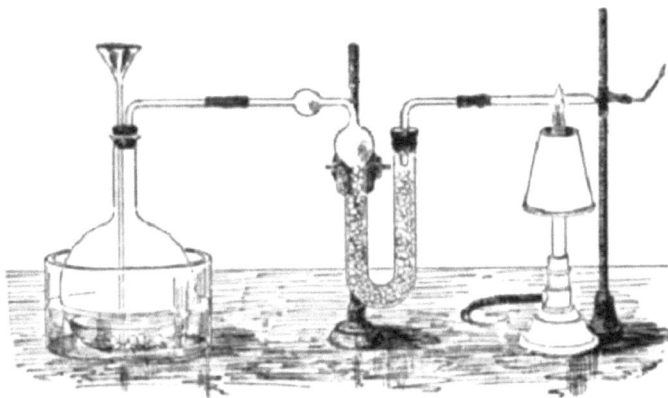

Fig. (7). A version of the Marsh test for arsenic (see text).

Hemlock and Death of Socrates

The death of Socrates is a classical example of poisoning by a plant substance, hemlock. Hemlock refers to poison hemlock, water hemlock, hemlock water dropwort, lesser hemlock (fool's parsley), and sometimes to other herbs. All are members of the same plant family that also includes edible vegetables including carrots, celery, dill, parsley and parsnips [25]. Indeed, an account from the University of Missouri, where I was appointed Professor for many years, states that "One of the first weeds that you can see 'greening-up' right now [right now referred to early spring] along roadsides, pastures, and a lot of other areas is poison hemlock (*Conium maculatum* L.)" [26]. This article states that poison hemlock produces many chemicals that are toxic to humans and other animals; the most toxic are the alkaloids coniine and Y-coniceine. Poison hemlock is a unique plant, the only one of this genus that produces the toxic substances known as alkaloids, according to Block [25].

The ancient Greeks instituted a form of capital punishment called "State Poison" the method by which the philosopher Socrates (c470 BC-399 BC) was executed by hemlock [27]. The last days of Socrates are described in Plato's *Phaedo* and are found in an accounting of this historic writing [28]. Socrates is offered escape by friends but insists the he cannot return evil for evil and has the duty to respect the law of the City that nurtured him. When the sacred ship arrives from Delos Socrates is unshackled and allowed a final visit with his sorrowing wife Xanthippe who arrives with their infant son. Socrates' final hours are spent discussing subjects including the immortality of the soul with friends (Fig. **8**).

The information that has come down to us indicates that the discussion included a justification of a life lived with a view to the cultivation of the soul. Plato's faith accepted the deathlessness of the soul and physical death that released the soul. Socrates addressed the whole group and is said to have smilingly said that the real Socrates would soon depart to the joys of the blessed and that only his body would remain to be buried. After washing his body (to free the women from this task), a final visit by his sons and womenfolk, and just before sunset a cup of hemlock was brought and Socrates drank it "… as if a libation to the Gods". His last word were said to be: "Crito, we owe a cock to Aesculapius; please pay it and don't let it pass." Aesculapius was the god of medicine.

Fig. (8). Death of Socrates by hemlock.

The account of Socrates' death informs us about the effects of hemlock, and is told in Plato's *Phaedo*, and summarized by Enid Block [25]. Plato describes a slow, ascending paralysis that started in Socrates' feet. It moved steadily up his legs to his chest. Socrates' mind remained clear until his death which arrived calmly and peacefully by Plato's account. Block [25] asks the question: whether this is, or could be, an accurate account, and says others have also questioned the calm death. It is claimed that hemlock poisoning would have produced a "far nastier and more violent end." Block [25], however, concludes after much research: "The calm, peaceful death of the Phaedo was an historical fact." He also raised the issue of what can be known for sure about the actual poison used. Block concluded [25]:

"In the end I have been able fully to align Plato's description with modern medical understanding. Socrates suffered a peripheral neuropathy, a toxin-induced condition resembling the Guillain-Barré syndrome, brought about by the alkaloids in *Conium maculatum*, the poison hemlock plant."

Like most of the story about Socrates' death, why he was to be poisoned by the state has been much debated. An interesting account [29] is provided in "Why Socrates Died: Dispelling the Myths.", an on-line review of a book by Robin Waterfield published in 2009. Below is a summarization of what I believe to be the essence of what Waterfield is saying. Athens, in 399 BC when Socrates was sentenced to death, was a radical democracy with freedom of speech a central government freedom. Why was Socrates sentenced to death when all he did was talk? This account describes the workings of the courts and that Socrates was convicted by a majority vote. Some background is useful for perspective. Athens in the late 5th century had waged a long war against Sparta and had been defeated. A typhoid epidemic had killed an estimated 25% of Athenians. There had been political coups (411 BC and 404 BC); democracy was restored in 403-402 BC but the society had been changed. Socrates was said to have had a love affair with Alcibiades and had a pedagogical association with Critias, who is considered to have been the most violent and outrageous anti-democrat of his time. By 399 BC both Critias and Alcibiades had been killed but Socrates was politically very active. Even Plato suggests that Socrates engineered his own death. I find it interesting that the trial of Socrates is described to be in the category of "assessed trials" in which the state acknowledged different degrees of guilt. The prosecutor of a guilty defendant would propose a penalty and the defendant could counter with a lesser penalty, after which there was another vote taken by the jury. Socrates, rather than proposing a fine or exile, is said to have made a mockery of this process by insisting that he wanted "free diners at public expense".

After his death, radicals known as cynics were said to have defecated in the streets; they performed other outrageous acts and were generally disorderly and unruly; they refused to consider themselves citizens; and they openly mocked the law [28, 29]. Does this remind you of anything today? *Plus ça change, plus c'est la même* (the more things change, the more they stay the same).

Death of Cleopatra by the Asp

I have chosen to use the story of the death of Cleopatra (c69 BC-30 BC) as a classical example of poisoning by venom. Cleopatra VII Philopator (known in history simply as Cleopatra) was the last active ruler of Ptolemaic Egypt. When her fleet of ships surrendered to Octavian's, Cleopatra rightly assumed that she was on the path to certain execution. She had prepared a mausoleum as a fitting place for her body to be placed after mummification [30]. I will not detail the tragic story (which has been told in many forms over the years) of the misunderstandings about each other's death prior to the actual death of her husband Marc Anthony who thought that Cleopatra was already dead and had fallen on his own sword. Cleopatra was under house arrest and refused to eat. Most ancient sources, according to the account referenced here [30], record that Cleopatra died following the bite of an asp which was smuggled into her chambers either in a basket or vase. Classical sources uniformly suggest that the snake bite was on her arm; it is popularly described that she clutched an asp to her breast (Fig. **9**).

Fig. (9). Death of Cleopatra by an asp.

It is somewhat puzzling that she would elect to die in this manner. Did she know that poisons that acted quickly were too painful and that the least painful poisons were too slow acting? Indeed, the bite of an asp is slow and very painful. However, it is generally reported that her death was peaceful and quick. Is this written to better fit the story line? The account I refer to here [30] says that many have suggested that the snake was actually a cobra. It is recorded that two handmaidens committed suicide with her. Cobras generally dispense all of their venom in a single bite so one snake would not have killed all three. Could three large cobras have been smuggled into her chambers? Mysteries, intrigue, and uncertainties abound in this story handed-down and written about by many, and retold in movies and plays, and even by Shakespeare.

NOTES

[1] Philosophical Dictionary, Voltaire. In the chapter: Enchantment. Magic, Conjuration, Sorcery, *etc*. https://ebooks.adelaide.edu.au/v/voltaire/dictionary/chapter178.html

² The Great Lozenge-Maker. A Hint to Paterfamilias. 1858, by John Leech (caricaturist (1817-1864), first published in Punch (London), 20 November, 1858 (In the Public Domain).

REFERENCES

[1] Arsenic: A Murderous History - Dartmouth Toxic Metals Superfund Research Program [Internet]. [cited 2017 May 22]. Available from: http://www.dartmouth.edu/~toxmetal/arsenic/history.html

[2] Albertus Magnus - New World Encyclopedia [Internet]. [cited 2017 May 6]. Available from: http://www.newworldencyclopedia.org/entry/Albertus_Magnus

[3] St. Albertus Magnus | German theologian, scientist, and philosopher | Britannica.com [Internet]. [cited 2017 May 6]. Available from:https://www.britannica.com/biography/Saint-Albertus-Magnus

[4] Arsenic – the "Poison of Kings" and the "Saviour of Syphilis" [Internet]. [cited 2017 May 5]. Available from: http://jmvh.org/article/arsenic-the-poison-of-kings-and-the-saviour-of-syphilis/

[5] LacusCurtius • Roman Law — Leges Corneliae (Smith's Dictionary, 1875) [Internet]. [cited 2017 May 6]. Available from: http://penelope.uchicago.edu/Thayer/E/Roman/Texts/secondary/SMIGRA*/Leges_Corneliae.html

[6] Faunce J. Lucrezia Borgia : a novel. Crown Publishers 2003; p. 277.

[7] Bradford S. Lucrezia Borgia : life, love and death in Renaissance Italy. Penguin Books 2005; p. 421.

[8] FOREIGN RELATIONS: Arsenic for the Ambassador - TIME [Internet]. [cited 2017 May 6]. Available from: http://content.time.com/time/magazine/article/0,9171,865337,00.html

[9] Comic cartoon about food adulteration, 1858, from Punch. The British Library; [cited 2017 May 22]; Available from: http://www.bl.uk/collection-items/comic-cartoon-about- food-adulteration-1858-f-om-punch

[10] Mr. Punch's Victorian Era: An Illustrated Chronicle of the Reign of Her ... - Punch (London, England) - Google Books [Internet]. [cited 2017 May 22]. Available from: https://books.google.com/books?id=tso-AAAAYAAJ&pg=PA293&lpg=PA293&dq=the+great+lozenge-maker&source=bl&ots=6TxH6nIZaA&sig=38zHKUpZphtQ_GAGnvKNbVIcEKE&hl=en&sa=X&ved=0ahUKEwiHnOydl4TUAhUoxFQKHWsUDpU4ChDoAQgxMAU#v=onepage&q=the

[11] Bradford Lozenge Poisoning Case - Geri Walton [Internet]. [cited 2017 May 22]. Available from: https://www.geriwalton.com/bradford-lozenge-poisoning-case/

[12] Murder By Poison | The New Yorker [Internet]. [cited 2017 Nov 8]. Available from: https://www.newyorker.com/magazine/2013/10/14/murder-by-poison

[13] Hempel S. The inheritor's powder : a tale of arsenic, murder, and the new forensic science 278.

[14] Schillace B. Bodle and LaFarge: Sensational Arsenic Cases – DITTRICK Museum Blog [Internet]. [cited 2017 Nov 8]. Available from: https://dittrickmuseumblog.com/2014/03/04/bodle-and-lafar-e-sensational-arsenic-cases/

[15] Lowndes MB. Great French Mysteries: The Strange Case of Marie Lafarge. McClure's Mag [Internet] 1912; 38: 603-18. Available from: https://babel.hathitrust.org/cgi/ssd?id=mdp.39015028799230;seq=641;number

[16] Madame Marie Lafarge, Murder, and Arsenic - Geri Walton [Internet]. [cited 2017 Nov 9]. Available from: https://www.geriwalton.com/madame-marie-lafarge/

[17] Jensen WB. The Marsh Test for Arsenic [Internet]. [cited 2017 Sep 11]. Available from: http://www.che.uc.edu/jensen/w

[18] Steinmaus C, Carrigan K, Kalman D, Atallah R, Yuan Y, Smith AH. Dietary intake and arsenic methylation in a U.S. population. Environ Health Perspect 2005; 113(9): 1153-9.

[http://dx.doi.org/10.1289/ehp.7907] [PMID: 16140620]

[19]   Singh AP, Goel RK, Kaur T. Mechanisms pertaining to arsenic toxicity. Toxicol Int [Internet] 2011; 18(2): 87-93. Available from: https://www.ncb.nlm.nih.gov/pmc/articles/PMC3183630/?report =printable
[http://dx.doi.org/10.4103/0971-6580.84258] [PMID: 21976811]

[20]   Hughes MF. Arsenic toxicity and potential mechanisms of action [Internet]. Toxicology Letters 2002; 133: 1-16. Available from: http://www.sciencedirect.com/science/article/pii/S037842740200084X
[http://dx.doi.org/10.1016/S0378-4274(02)00084-X] [PMID: 12076506]

[21]   Rosano TG. Ellenhorn's Medical Toxicology: Diagnosis and Treatment of Human Poisoning Matthew J Ellenhorn, Seth Schonwald, Gary Ordog, and Jonathan Wasserberger 1997; 2047 pp.. $199, ISBN 0–683-30031–8. Clin Chem [Internet]. 1998 [cited 2017 May 22];44(2). Available from: http://clinchem.aaccjnls.org/content/44/2/366

[22]   Ratnaike RN. Acute and chronic arsenic toxicity. Postgrad Med J 2003; 79(933): 391-6.
[http://dx.doi.org/10.1136/pmj.79.933.391] [PMID: 12897217]

[23]   Gehle K, Harkins D, Johnson D, Rosales-Guevara L, Dennis-Flagler D, Drehobl P, *et al.* Case Studies in Environmental Medicine ATSDR/DTEM Revision Content Experts. [cited 2017 Nov 9]; Available from: https://www.atsdr.cdc.gov/HEC/CSEM/arsenic/docs/arsenic.pdf

[24]   Oremland RS, Stolz JF. The Ecology of Arsenic. Source Sci New Ser [Internet] , 2003 [cited 2017 May 22];3002069(5621): 939-44. Available from: http://www.jstor.org/stable/3833922

[25]   Bloch E. Hemlock Poisoning and the Death of Socrates: Did Plato Tell the Truth? Oxford University Press , 2001 [cited 2017 May 6]; Available from: http://users.manchester.edu/Facstaff/SSNaragon/ Online/texts/316/Bloch

[26]   Spaunhorst D, Bradley K. 2012. Weed of the Month: Poison Hemlock [Internet]. Food and Chemical Toxicology. 2012 [cited 2017 May 6]. Available from: https://ipm.missouri.edu/IPCM/2012/Weed-o- -the-Month-Poison_Hemlock/

[27]   h2g2 - A Brief History of Poisoning [Internet]. [cited 2017 May 5]. Available from: https://www.h2g2.com/approved_entry/A4350755

[28]   Phaedo - Socrates death by Hemlock Poisoning [Internet]. [cited 2017 May 6]. Available from: http://www.age-of-the-sage.org/greek/philosopher/phaedo.html

[29]   Waterfield R. Why Socrates died : dispelling the myths [Internet] Faber and Faber , 2009 [cited 2017 May 6]; Available from: https://www.timeshighereducation.com/books/why-socrates-died-dispell- ng-the-myths/406864

[30]   Pharaohs of Ancient Egypt: Cleopatra VII [Internet]. [cited 2017 May 6]. Available from: http://www.ancientegyptonline.co.uk/Cleopatra-death.html

CHAPTER 3

Scorpion Venoms

Abstract: More than 1500 species of scorpions are known and they populate every continent except Antarctica. They abound in dry and desert areas of the world. By most accounts, only about 20 species have venom that can kill a human. Mostly, they live secretive lives preferring underground burrows or crevices in rocks by day and come out at night when they hunt for prey that for different species ranges from insects to rats, mice, lizards and other small animals. Scorpions produce complex venoms composed of many chemicals and they possess the ability to control the amount of venom injected by a sting, and even control the quality of that venom. The most toxic venom for humans appears to be that produced by the death stalker, and there are four chemical components. The earliest effect is on the transmission of nerve impulses which creates paralysis, and a subsequent toxicity to the heart. The most significant nerve toxicity is described as an impairment of calcium movement that blocks transmission of impulses across synaptic junctions to cause paralysis.

"And they had tails like unto scorpions, and there were stings in these tails; and there power was to hurt men five months."[1]

Keywords: *Androctonus crassicauda*, Antivenom, Autonomic nervous system, biotoxin, Calcium channel, CNS, death stalker, Fat tail scorpion, *Leiurus quinquestriatus*, Man killer, Myoneural junction, Nervous system, Neurotransmitter, Postsynaptic neuron, Presynaptic neuron, Scorpion, Synapses, Venom, Voltage-gated channels.

INTRODUCTION

Venoms are a type of biotoxins which are designed to be injected by specialized hollow fangs or by stingers, and are found in some insects and arachnids (spiders and scorpions). Other poisonous substances are found in some plants, in microorganisms, and in snake venoms. The reason for the existence of these poisons is sometimes apparent and sometimes mysterious. I shall focus on venoms from two scorpions, one known as the 'the death stalker' (Fig. **1**) and one called the 'fat-tailed scorpion' (Fig. **2**). The death stalker is widely known as the most deadly to humans of all scorpions. The literature affirms that there may be as many as 2,000 known species (we don't know about those we don't know about); however, only about 20 different species are venomous enough to put an adult human at risk of dying [1]. The deaths talker leads the list. It is not particularly ominous in appearance or size but its venom is a witch's brew of neurotoxins capable of allowing the scorpion to successfully prey on small mammals such as desert rats many times its size. The death stalker (scientific name: *Leiurus quinquestriatus*) is found throughout desert regions in the mid-east and in northern Africa. The name translates approximately as "five-striped smooth-tail". It is nocturnal and spends daylight hours underground or beneath rocks, to escape both light and heat. It appears that envenomation of humans is accidental when the scorpion is in danger of being crushed.

Fig. (1). The death stalker scorpion, *Leiurus quinquestriatus.*

The fat-tailed scorpion (scientific name: *Androctonus crassicauda*) is larger (up to about 4 inches in length), brownish colored with reddish claws and as the name implies it has a large tail (which contains a deadly stinger). The genus name, *Androctonus*, is Latin and originates from the Greek and is translated as 'man killer' and is said to cause several human deaths each year [2].

These species, like all scorpions, take their food only in liquid form. To feed on an animal such as a rat, the scorpion must incapacitate its prey quickly. Then, as disgusting as it sounds, using powerful appendages and mouthparts, bits of flesh are removed and taken into a pouch where they are partially digested and then sucked into the scorpion's internal digestive system.

The short story to follow is the way I have decided to illustrate scorpion envenomation. It should be kept in mind that toxic agents never exert their effects on a blank slate; every human being reacts somewhat differently. Also, a spoonful of humor helps the science along.

Fig. (2). The fat tail scorpion, *Androctonus crassicauda.*

THE DAWN SINGER

The professor paused in the scant shade of a low, rocky outcrop jutting from the barren, alien landscape. He was a small, wiry man with bandy legs, thin brown hair and piercing eyes. Reaching up, he pushed back a frayed pith helmet on the front of which could be seen in faded letters 'Speculatus' (Fig. **3**). Heat and a throbbing, classical migraine had drained his will and he was flooded with depression that spiraled downward toward despair.

Removing gold wire-rimmed glasses, Speculatus mopped his brow with a large red bandana. He blinked rapidly from the stinging perspiration draining in rivulets from under his helmet in spite of his attempts with the now sweat-saturated bandana.

The solitude of the desert caused him to reflect on his situation. He was alone, removed from civilization, far from medical help, and surrounded by certain and unique dangers peculiar to him. Since childhood, he had been acutely aware that he suffered from classical migraines. Attacks always came unexpectedly, began gratuitously with an aura, and proceeded to tunnel vision accompanied by a deep feeling of panic and impending doom. Awareness of this caused him to be more introverted and also to realize that he could become partly incapacitated at any time with only a brief warning. Lately, his symptoms had increased in frequency and severity and had bordered on near-hallucinations which were sometimes triggered by certain sound frequencies, when repeated.

Fig. (3). Speculatus. Original art by John Allen.

Replacing his glasses, Speculatus hunched forward and ran his thumbs under the straps of the large pack strapped to his back. Attached to the pack and inside numerous pouches on it, could be seen the outline of odd-shaped equipment. He pushed upwards on the straps while hunching his shoulders to momentarily lift and shift the load. A slight breeze gained access to his sweaty shirt back, and momentarily brought relief. Then he resettled the load, and the strain returned.

Professor Speculatus had a brilliant mind and he read widely in all the basic medical sciences. He was not a hypochondriac, but he was acutely aware of his body and the signals it sent, and rarely went to physicians, but often self-diagnosed. His migraines were of a special type quite unlike the garden-variety migraine experienced by most migraine suffers. Ordinary migraines produce terrific one-sided headache and nausea that can become totally incapacitating. The migraines he suffered originally were called "classical" migraines in the older literature, but later were labeled as "migraine with aura". They produced a bewildering array of symptoms, but generally not the characteristic one-sided headache. Typically an aura and strange feelings preceded an attack, which might be brief and simple, or long and complex. Speculatus' attacks always began with an aura. His central vision suddenly disappeared, replaced by a jagged hole in the center of his visual field. This proceeded to tunnel vision and was accompanied by a deep feeling of panic and of impending doom. This medical warning of neurological difficulty began with sensitivity to sound, light and touch, and then progressed to a bewildering aphasia (out of synchrony) of multiple senses.

At its worst, during an episode, Speculatus became aware that he could not access any memory, long or short-term. Aware of this, he would attempt to test his limitations and could not bring up the name of a close friend, or even his mother's name. Words would not form into logical sentences, but he was vividly aware, as in a lucid dreaming state. Unable to see clearly, with distorted sensory input and no memory or ability to organize thoughts, it had produced near panic when as a child he first experienced a full attack. He had learned by experience that the phases generally lasted only minutes and one would abate as another initiated. As vision and memory returned numbness and tingling of the fingers of his hand would bring thoughts of a stroke, but blood vessels feeding specific regions of the brain were experiencing contractions and dilations, rather than clots or hemorrhage. Fortunately, there had been no permanent effects, but sometimes a delayed headache struck so profoundly as to immobilize and require withdrawal from all light and sound. By the following day, after sleep, he might experience unexplainable pain in any part of his body, but paradoxically he often experienced days of extraordinary creativity and elation. Lately, however, Speculatus had experienced worsening depression and symptoms had developed that he feared were bordering on schizophrenia.

Speculatus clinched his eyes and pressed the thumb and first two fingers of his right hand against his throbbing temples near his eye sockets. He shifted the pressure to his eyeballs and a pattern of white, then colored, lights flooded his internal vision and then receded as he relaxed his hand. Was an aura imminent? His reflections deepened. He had money for only a few more days of field work. His life's work appeared to have come to nothing. How could he ever fund another field expedition? His theories now would never be proven.

Suddenly, a green bottle fly (commonly known as the blowfly) droned in and, unexpectedly but with directed purpose, lit at the corner of his mouth. He felt it's disgusting feet, which had so recently rested on rotting carrion, on his lip and struck at it reflexively. The fly evaded the blow and demonstrated the Doppler Effect[2], hypnotically buzzing near his ear and then receding with a pitch that rose and fell, rose and fell. The buzzing synchronized with the throbbing in his head as Speculatus focused a gaze so piercing that it seemed capable of striking the fly to the desert floor.

Confusing thoughts buzzed and throbbed in his brain in rhythm with the loud, drone of the fly which continued to approach and recede. Images of his long days of futile searching in the desert continued to evolve in detail. During these times of introspection, which had come with greater frequency lately, what was real and unreal in this unreal, lonely land sometimes became impossible for Speculatus to separate. Yet, somewhere deep inside, hope of a great discovery battled to remain alive.

Speculatus was not aware of the time that passed as his strange mental state deepened. Then the buzzing in his brain temporarily receded and was replaced by panicky thoughts of the dangers he faced alone in the desert. His mind focused to thoughts of venomous desert creatures and vividly created the image of the scorpion known as the death stalker.

At this moment, in a deep, shaded crevice near the base of the rock on which Speculatus leaned there was a stirring – a death stalker scorpion awakened (Fig. **4**). The central part of the scorpion's nervous system, not quite a brain, was fully capable of coordinating the scorpion's movements as it directed its vision down the crevice toward the outside world. Bundles of nerves ran to powerful pincers, six legs and its central offensive device for obtaining food and for defense – a lance-like stinger that was connected to a venom sack controlled by strong muscles. If the primitive brain of the creature had been capable of knowing, it would have understood that its biochemistry was capable of producing four distinct toxic chemicals that could quickly immobilize the nervous systems of crickets, mice, rats, and even humans. Although not designed for large animals, humans have died from the sting of this scorpion. Crickets and mice are almost instantly paralyzed by the venom. If the scorpion had been cognizant, it would have known that only 5 micrograms of its venom was enough to kill a mouse.[3] Five micrograms is equal in weight to about 1/100th of a grain of table salt. About 18 milligrams (a milligram is 1,000 micrograms) would likely kill an envenomated human. This is only an estimate; obviously, no controlled laboratory experiments have been done. A tenth of this dose could be lethal for the most sensitive persons or a child, and a large scorpion that had conserved its venom, cold inject this amount. Indeed, the scorpion usually tries to conserve its venom and it has powerful striated muscles controlling the stinger and associated venom sacs. The scorpion can choose to inject from practically no venom or all of its supply which could be several hundred micrograms.

Fig. (4). The desert outcropping of ancient rock in the desert.

The outside world was still sunlit and the death stalker sensed this and retreated. It preferred the cool, dark depths inside the rock crevice in the unusual rock formation against which Speculatus leaned.

Speculatus slowly became aware that the needle of the ordinary compass attached to his belt began to fluctuate very slightly and periodically. The compass sensed a very faint and unusual magnetic field and this was communicated to the electronic instruments secured to his backpack. Speculatus was slow to realize this because of his mental state and he leaned back against the only smooth area of the outcrop, letting the rock support his weight. The physical relief which this brought did not extend to his mental state which continued to be filled with anxiety.

Then, suddenly as had begun to happen more frequently of late, Speculatus knew that a serious, full-blown migraine attack had begun. The first sign, as always, was the appearance of colored, zigzag lines and an irregular-shaped hole in his visual field that moved with his gaze to obscure any object near his central focus. This was accompanied by a deep feeling of panic and impending doom. Speculatus had learned to control this feeling, through much effort of will. He covered his eyes to block the light and remained, strangely, immobile and leaning against the rock.

A dust devil arose in the nearby sand, danced eerily, and swirled directly up to the rock face. Its actions intruded into Speculatus' consciousness. With a sensation experienced only by those who, like Speculatus, are completely alone, this inanimate whirlwind seemed to detect his presence and purposefully to seek him out. In a final burst of energy, the hot swirling wind of the dust devil spent itself against the stone face on which he leaned, sending bits of sand and debris stinging against his face and hands. Speculatus held his breath and reflexively closed his eyes until the desert quiet returned.

Sometime later, Speculatus roused himself. His head, felt strange, and he was slightly nauseous, but his vision disturbance (aura) had mostly passed. This time, maybe, he would be lucky.

Gradually, Speculatus became aware, then alarmed, as the realization sank in, that for some time he had been sitting on the sand and rocks, his body a possible source of warmth for cold-blooded desert vermin as night approached. Speculatus, already generally knowledgeable, had read avidly before journeying here, about desert plants and animals, and implicitly about the poisonous scorpions of the region.

As Speculatus grew more aware of his surroundings, he was visibly alarmed as he observed the sun's angle and saw that the blood-red disk was half-eclipsed at the horizon. Anxiously, he realized that it was well-past time that he should have begun the journey back to his base camp. He had no desire to be caught in the desert after sundown. Indeed, Speculatus' fears of the desert at night were not entirely irrational. But, like the other fears that frequently visited him as unwelcome guests, he managed to control the rising panic, but only with great conscious effort.

Then, impulsively, he decided to make one more survey with his instruments, and after inspecting the area nearby, he selected the rock against which he had recently rested. Hurriedly, Speculatus unhooked earphones from one side of his backpack, and slipping them on he stepped a few paces away from the rocky outcrop. Unfolding a small but sensitive antenna, he held it horizontally and rotated it slowly in an arc. The antenna fed weak signals from the surroundings through a powerful laptop computer to an analyzer with an output, some of which went to the earphones.

The equipment was unique and had been entirely designed and assembled by Speculatus. Suddenly, he thought he heard a faint buzzing, or was he imagining it, or was it another part of the migraine aura? Sinking down to his knees, Speculatus stabilized himself and focused his attention on his strange instrument whose dials he began to adjust. He appeared motionless, head drooped, eyes closed and shoulders hunched. The sun sank lower, and disappeared.

The death stalker, inside the crevice, suddenly alerted, sensing alarming danger. Deftly manipulating its six legs, it whirled in the narrow passage in the rock to come pincer to pincer with another scorpion— the fat-tailed desert scorpion (scientific name: *Androctonus crassicauda*[4]) who vied for this ecological niche. The fat-tailed scorpion was larger; four inches in length and dark brown with large reddish pincers and a large fat tail with a prominent stinger it the tip. It carried toxins in its sting that were the near equal of the death-stalker and it was in no mood to share territory. It too controlled when and how much of the precious poison to dispense in any given sting, and it was aroused and as close to anger as possible with no real brain. The combination of chemicals stored in its venom sac included chemicals scientifically known as calcium channel blockers. Translated, this simply meant that they cause increased transmission of impulses from the central nervous system to muscles which contract continually and excessively, resulting in paralysis of its prey.

The two scorpions in quick succession tumbled out of the fissure in the rock and contacted the collar of Speculatus' shirt and immediately both engaged pincers, with the larger fat-tailed scorpion having a big advantage. They both sought refuge inside the dark opening that was the back of the collar of his shirt.

Speculatus flinched. Was he now imagining it or was something crawling along on his neck beneath his collar? Then, all doubt was gone as a piercing pain struck at the flesh of Speculatus' neck; and then came another blinding pain. Flailing his arms he stood erect and gasped as the pain that felt like nothing he had ever experienced. Reaching back, he felt, then grasped, the entangled scorpions, which he flung from his body onto the sand at his feet. The scorpions disengaged and he instantly recognized the desert, yellow scorpion, *Leiurus quinquestriatus*, sometimes called the "death stalker", and knew its deadly venom was already in his body. A flood of emotions−disgust, revulsion, and fear−washed disturbingly over him.

The death stalker curled its tail over its back, raised its claws menacingly, and waggled itself like a miniature Go-Go dancer. With disgust, Speculatus ground at the scorpion with his foot. Raising his boot, Speculatus saw the scorpion's hard outer skeleton had protected it well against the soft sand and it was still full of fight. Speculatus quickly pushed the scorpion with the toe of his boot, and ground it with his boot heel against a rock, even as it continued to rear its pincers and tail erect. He then saw the fat-tail scorpion scuttling into the rock crevice. With awful pain, he began to extricate himself from the heavy pack of instruments. After carefully examining the site for other scorpions, he sank back against a smooth protrusion of the rocky outcrop, emotionally and physically exhausted.

Speculatus reached back and lightly ran his fingers over the area of his neck that already had begun to swell as the searing pain stabbed repeatedly into his flesh. The initial effect of the venoms was searing pain from the witches' brew of chemicals present in each. He tried to think analytically. The lingering aspects of the migraine aura compounded by the scorpion toxins made this difficult, but he was grateful that his cognition and memory still functioned. Both toxins were similar and their effects would be additive; they might even be compounded. The consequences to his physical state depended on how much venom had been injected, and with two envenomations close to his brain, he feared the worst. Had anyone ever been envenomated by two deadly scorpions at the same time? He recalled that within three-quarters of an hour, the venom would cause the hyper-adrenergic state[5] with symptoms that included rapid and irregular heart beat. This was a fancy description of severe interference with the automatic control of essential bodily functions by the brain. Within an hour and a half he could be quite ill. He immediately recalled the toxicological definition "SLUD" used since graduate school days to mnemonically recall the symptoms of envenomation by scorpions such as these. He would have intense salivation, lachrymation and incontinence of both types. Subsequently, he might have severe neurological symptoms including convulsions. He would experience severe muscle weakness and ataxia and could have lung edema and cardiac collapse and even death. What he feared most were the expected neurological effects. Of the four main toxins in the venom cocktail, one was even now disrupting the electrical signals of nerve transmissions to his muscles. Most significantly, the control to his diaphragm that controlled breathing was compromised. He wondered how this venom might affect his mental condition. Had anyone ever suffered a sting from the death stalker, so close to the brain, while in the throes of classical migraine? What might be the synergistic, added consequences of two scorpion stings? Was he merely being analytical and borrowing trouble, to make this involved analysis? Hopefully, the visual impairment of the aura seemed to be growing less.

As these clinical considerations flooded his mind, he considered the odds. Believing in preparedness, he had scorpion anti-venom at his base camp. However, it was experimental and might not be fully effective. At least, it was sold regionally and was likely to have been prepared against the local sub-species, he reasoned. He reflected on the distance to his camp; it was at least two hours away, by the most direct route. He knew the neurological symptoms of the venom would strike first, and the effects on breathing would be most life threatening but the impairment of his leg muscles could immobilize him. He feared most the cardiac damage that might occur later. The exertion of reaching camp by foot would make the symptoms worse. More frighteningly, he vividly imagined his vulnerability to a lion or a pack of disgusting wild dogs, or hyenas, if he should convulse somewhere out in the dark desert. By remaining where he was, he would have no anti-venom, but would he have fewer symptoms by resting, and less risk from unknowns in the desert?

He assessed his condition. He had the heavy pack which contained his instruments. His first impulse was to carry his precious instruments out with him. He began to slip his arms through the pack straps and the pain immediately seared into his neck. Carefully, he extricated himself and clasped his neck as the pain crescendoed and then abated slightly. Examining the pack he inspected its contents, hopefully. Besides the special instruments, its contents were one flare, some matches, a canteen partly filled with water, and a short-barreled, thirty-eight caliber revolver, loaded with five cartridges. He drank from the canteen, tucked the flair and matches into a pants' pocket and hefted the revolver. Considering his expected rapidly weakening state he decided against, then for, the pistol and tucked it into his belt.

His logic was then abruptly interrupted by strange animal sounds which crystallized his decision to leave immediately. He quickly collected his instrument pack, and reflecting that this was an isolated spot and that his pack contained no food to attract animals, he decided to place it on a low rock that jutted from the main rocky outcrop. He forced himself to drink the last of the water from his canteen and laid it beside the pack.

Introspectively, he sensed that his pulse already was quickening and then he felt a slight, transient twitching of the left side of his face. The first neurological sign of the venom's action, he realized. He stood erect with effort and took his bearings. He followed with his eyes the direction of travel he had taken into this desert spot. He reflected, however, that his path to the spot had been circuitous and he began to consider where he might take short cuts. His weeks spent in the area had given him a good familiarity with the region. However, today's trip had extended into an area he had not explored before. It was rugged terrain and he would be concerned about finding his way back to camp, even with all his faculties. Later, as the venom's biological action grew worse, as he knew would happen, would he be able to both struggle on and keep to the correct direction? Unfortunately, his decision to try was not revocable.

Summoning all his courage, he began to retrace his desert footprints which were already being covered over by the sand carried by the swirling, intermittent wind, like confectioner's sugar on a dessert. Even in his state, the pun did not escape him. As he struggled to walk as rapidly as possible, the pain in his neck increased. Then, he felt the neurological symptoms worsen, and to physically wash through his brain and simultaneously warp his consciousness and he feared he would lose consciousness. His vision blurred and he could not think clearly. Was this the venom, or was another aura beginning? His heart rate and respiration now were unnaturally and frighteningly rapid and still increasing. Drool came from his lips and his eyes watered copiously. He felt an intense need to urinate. These were symptoms of the powerful toxin at work on the nerve synapses that connected his automatic nervous system with his muscles.

Speculatus suddenly became aware of a shadowy form to his left. His vision was bleary and unclear, and his central vision was almost totally gone from the migraine aura. It was now growing quite dark, because blackness descended, without twilight, in the clear, desert air as soon as the sun fell below the horizon's rim. Trying to focus his vision was an intense effort and, at first, he could not tell if there were other forms joining the shifting shadows. Then he was sure--- two burning eyes stared at him from the darkness; or, was this hallucination? He paused to gain his balance and consider his options. Merely thinking was an effort; rational thought a near impossibility. It was curious to be conscious of making an effort to try to think, a normally unobserved process for his brilliant mind.

Suddenly he realized that the fasciculations of his cheek were starting again, and now grew more severe. This twitching of the small muscles of his face was a premonition of greater weakness to come in his leg muscles and of possible convulsions and muscle paralysis. He recognized this as the effects of the toxin called chlorotoxin in the scorpion venom. He contemplated his predicament and fear rose up inside. He was not capable of sound judgment, but realized in more lucid moments that his convulsing and subsequently unconscious body would become prey for the forms in the darkness which he feared were hyenas, or possibly, lions. Feeling great weakness, he sank to the sand and then attempted to roll onto his back as shadowy forms shifted several yards away. Fearing he might convulse or collapse, he was temporarily overcome by irrationality and drew his pistol and emptied it into the dark forms that snarled, yelped and seemed to disappear.

Feeling too weak to flee, he decided to hope his symptoms might abate. Temporarily, his mind did seem to clear slightly, but to his dismay, the shadowy forms and glowing eyes were still there. He reached into his pack and withdrew the flare; the one device he had left for defense. He lit the flair and as it popped and glowed its brightness drove back both the darkness and the dark shadows which retreated with the temporarily advancing edge of light from the flare. He managed to shove the end of the flare into a dry, dead bush which flamed up and persisted as the flair began to fail. Speculatus drug additional brush into the fire. This exertion increased his symptoms and suddenly he jerked with a grand mal seizure - an extreme effect of the scorpion toxins. This violent reaction brought him to the ground and his legs and arms alternately flailed into and through the small, but growing fire, and he lost consciousness. His clothing smoldered but did not flame up. After several convulsive jerks, his body grew still.

After a time, Speculatus aroused. As do all who survive a convulsion, he had no memory of the convulsive spasms, and their associated pain. He made a great effort to sit up and felt worse than he ever had in his life. He ached all over; even his hair seemed to hurt. His back and legs were throbbing from the violent contractions of the convulsions. His left sleeve was charred from the smoldering fire, and this was inexplicable, since he did not recall falling into the fire. He was frightened by the prospect of the return of the eyes and shadows lurking in the darkness, which he did selectively recall. He continued to tear and drool uncontrollably; he urinated with incontinence. His heart was beating at a frightening pace and his breath came in rapid gasps and he had to forcefully breathe. Most frighteningly, his mind seemed about to leave his body and no thoughts could be marshaled or controlled. He huddled near the fire and an unmeasured amount of time passed.

Sometime later, Speculatus' mind cleared somewhat and he sensed that his breathing and heart rate had abated slightly. The fire was only smoldering embers and, alarmed, Speculatus managed to struggle to his knees. Fortunately, the moon now had risen, and its strengthening light matched his improving mood. From the moon's position, he realized that minutes, not hours, had passed. Using dimly visible horizon cues, he began to move again in the direction he hoped was toward his base camp. He was weak and his gait was slow but by struggling he moved one leg in front of the other.

Blessedly, he began to recognize terrain features and then his camp became visible below, in the light of the gibbous moon now riding higher in the sky. As Speculatus staggered into his tent, he began to feel huge relief; then he suddenly convulsed again.

Later, he became aware that he was lying on the floor of his tent. He struggled to his feet and searching his almost inaccessible memory he recalled where he had stashed his supply of freeze-dried antivenom. He had purchased the antivenom from a local source and was aware that, unlike antivenom produced in the United States, it was not approved by any agency such as the Food and Drug Administraton[6]. However, being produced locally, he logically concluded it would be most effective against scorpions such as the death stalker and the fat tailed scorpion. Lacking a source of refrigeration, he had freeze-dried the antivenom in his laboratory. He hoped that this would help preserve the activity of the proteins that were the antibody against the toxin. He dissolved the freeze-dried antivenom in an intravenous injection bag of sterile saline which he hastily attached to a needle which he thrust into a vein in the back of his hand and secured it with surgical tape. He hung the bag above his bed, connected the intravenous drip and dropping down onto his bed he sank deeply into a state that combined delirium, stupor, and sleep.

Later, Speculatus roused. He disconnected the now empty intravenous bag. His neck was stiff and swollen and as he attempted to examine the sites of envenomations with his fingers, the swollen area throbbed incessantly. He drenched the wound with hydrogen peroxide, and applied antibiotic and a loose bandage. Then, he attended to his burns which were not serious. He was aware that within twelve hours he would possibly die or the major serious effects of scorpion venom would pass but he would be left with weakness and malaise. He checked his watch and calculated that more than 12 hours had passed since the scorpion stings. Soon, he might be over the worst risk of fatal cardiovascular or neurological effects.

Speculatus forced himself to consider breakfast. He must regain his strength. His camp was modest. It consisted of a tent, a small supply of food and water and some scientific equipment. He had no cell phone or other means of communication. He had accepted the great risk of being alone because of his limited finances and his personal traits of risk-taking. He had no close colleagues and he took little though about his personal safety. He had brought his supplies and equipment in by an ancient, rented Land Rover. Speculatus mentally reviewed his planned return date and realized he had to leave within three days if he was to keep his commitment to the college where he held an adjunct assistant professorship.

He forced down a pint of water and after lighting a primus stove he prepared and ate a breakfast of powdered eggs, bread, tinned ham and hot tea. He managed this activity by moving his head as little as possible. After eating, he felt somewhat stronger and slept fitfully until late evening.

As darkness approached, he prepared a hasty meal; ate sparingly without appetite and again slept fitfully through the night. Awakening in late morning, he was drenched in sweat and extremely thirsty. After drinking water and some tea, he again felt somewhat stronger. His thoughts returned to his equipment and to the hopeful buzzing he had heard - or had he imagined that - just before the sting by the scorpions. The mental image of the scorpions was revulsive. His clearing mind now was flooded with the excitement of the barely-recalled faint signals from his equipment, and most importantly, what it indicated! He must return immediately to the site. His rising emotion was then overwhelmed by anxiety and concern about having left his back pack of precious equipment unattended. The excitement of his potential find, the impending date of his leaving the area, and concern for his physical state swirled in his mind, which vacillated between extraordinary logical order, and strange unreality.

He resolved to return immediately to the desert outcrop. He packed food for two days, fresh equipment batteries, reloaded his pistol, and filled a large canteen of water. Then, as an afterthought, he attached a flashlight to his belt. The combined weight was not great but he was still stiff, sore, and in a weakened state, and his travel was slow.

The route back across the desert floor was vague in his mind and he strayed, and again found his way forward many times. It was past mid-day when the outcrop became visible. He was drenched in sweat and his legs felt weak from residual effects of the venom. He reflected that it had been miraculous last night that he had found his way out, while sick and in near darkness. Anxiety and concern mounted within him as he approached where he had left his pack and other equipment. It was not there! Then, with relief, he realized that the site he had chosen was a little farther on. He was immensely relieved to find his equipment undisturbed except for a covering of dust.

Shaking the pack carefully free of sand and possible scorpions - none were found - he replaced the batteries, eased it onto his back, and adjusted the dials. The upper body exertion, the heat, and the anxiety blurred his vision. He was unsteady, and leaned against the outcrop for support. His mind refused to function properly. Speculatus drifted in and out of reality.

Gradually he became conscious that a green bottle fly- perhaps the same disgusting fly as yesterday- had arrived to buzz annoyingly near his ear. The heat, the sudden loud buzzing, his throbbing head, and the tiredness that pervaded his body began to drain his will. As had happened ever more frequently in recent weeks, and greatly so since the scorpion's envenomation, an aura threatened and Speculatus' head felt odd, and he was flooded with anxiety and depression that deepened and spiraled downward toward black despair. The synchrony of the buzzing fly, in the otherwise silent landscape, pierced into his brain and in vain he sought some mental escape. No productive thoughts could be connected and he gave up the attempt. His stare followed its buzzing flight, back, and forth, back, and forth. The buzzing fly and the throbbing in his head locked in synchrony.

Speculatus momentarily drifted back closer to reality. He thought about the year he had spent in this desolate place, a long year but insufficient time to become accustomed to the outrage of having his person invaded at will by disgusting vermin such as the scorpion, and now, the fly - again.

He sank to a seated position at the base of the rock, and his reflections deepened. He had money for only a few more days of field work. His life's work appeared to have come to nothing. How could he ever fund another field expedition? The last chance to prove his theories were vanishing like morning moisture on the desert sand.

Images of his long days of futile searching continued to evolve in detail. During these times of introspection, which had come with greater and greater frequency lately, what was real and unreal in this unreal, lonely land became impossible for Speculatus to separate. Yet, somewhere deep inside, hope of a great discovery battled to remain alive and Speculatus was unaware of the time that passed as his strange mental state deepened.

The Doppler shift buzzing of the green bottle fly and the buzzing in his brain continued in synchrony and suddenly was joined by a buzzing from the instrument pack on his back. Trance-like, Speculatus grabbed his instruments and, because malfunctions sometimes occurred, began to check the wires and switches that connected the several components of the complex device. The instrument buzzing grew stronger and its synchrony intruded into, and dominated, his brain function. Speculatus' physical movements became odd, as if in a trance, as the instrument's buzzing ebbed and strengthened with the movement of the antenna which Speculatus haltingly rotated. The antenna pointed directly to the rock face where he had rested moments before.

Speculatus dragged the earphones from his head, shutting out the sound. He roused somewhat and only then did the excitement of his success begin to register. Could it be true! He sank to a stable position on his knees, unmindful of the possibility of scorpions. He replaced the headphones and adjusted the instrument repeatedly to examine the nature of the buzzing sound.

Speculatus then excitedly arose, and stiffly made his way to the rock face and inspected it. Using a geologist's hammer, he chipped several small rock samples from the rock face and examined them closely by eye and with a magnifying glass. Then he began to circle the rock formation in the rapidly failing light.

Suddenly Speculatus saw an almost hidden, narrow, long crack leading into the rock near its base and progressing into the overhanging rock. On closer inspection, the stone appeared to be quite weathered at this site.

Speculatus pried at the crack with his hammer and suddenly the rock above shifted and rubble showered down around him. By removing more loose stones, the professor began uncovering a narrow passageway leading into the rock itself. After considerable labor, a passage deep into and beneath the rock was revealed. Past the outer crack, the natural opening immediately was wider. After squeezing inside, Speculatus could stand erect. Using the flashlight, he saw that the passageway was blocked immediately by an unusual vertical rock formation. As if expecting this, Speculatus produced three small but complex microphones which he mounted in a triangle flush against the wall. He then plugged them into the backpack equipment. Speculatus next produced a miniature but powerful laser from a mounting beneath his backpack. Using the laser which completely drained his main battery, he then cleaved the rock and recorded the unique energy which escaped from beyond the rock wall through the lased fissure. Speculatus continued for some time to make observations and measurements within the cave.

At last Speculatus paused from his work. As his mind began to explore his feelings, he was overcome with emotion. The vindication of his theory, the satisfaction of proving his detractors wrong and the enormity of his discovery combined to produce euphoria. Another man might have leapt about or shouted, but the professor simply relished these thoughts deep inside.

An unidentified sound, disturbingly loud, beastly and nearby, jarred into the consciousness of Speculatus. Suddenly, he became intensely aware again of his present surroundings with its possible dangers. His discoveries had consumed the remaining daylight! With the return of awareness of his surroundings, Speculatus was distraught that he was again in the desert at night. He quickly collected his equipment and began to retrace his steps back toward camp. The moon had risen slightly earlier than on the previous night and his flashlight aided him. The shadows again appeared, and frightened and hurried him on his way. No eyes appeared, however, and he returned uneventfully, but with considerable pain, to his base camp. He then made his way to the airport and took the first available flight back to the United States. Within days, Speculatus was back at his college laboratory where, incessantly plagued by worsening migraines, he drove himself relentlessly to analyze the sound he had recorded.

Speculatus' work had previously been ignored or barely tolerated by his colleagues. All this changed dramatically when he was contacted for an interview by a young journalism student from the local press. The story was front page news in the college paper. The national news services picked it up and the professor was inundated by calls from the national media.

NEW FOSSIL WOMAN, EOPITHECUS OPERAII, DISCOVERED proclaimed the headline in a metropolitan newspaper. The lead paragraph began: A dramatic finding has just been released by Professor Ira Speculatus following a year in Africa's Afar region. The professor claims to have recorded a 29 million-year-old fossilized human voice. Speculatus says this vindicates his theory that sound can be fossilized. Furthermore, he says it proves that human evolution occurred much earlier than previously thought, since the fossil voice is fully human.

Speculatus' life, which was never normal or stable before these events, took a precipitous turn toward the bizarre. He had survived only because his college, like most great centers of higher learning, managed to tolerate a few odd individualists so long as they didn't overtly embarrass their peers. Speculatus' unusual work had been largely ignored and, therefore, tolerated.

As a direct result of this new and unwelcome publicity, Speculatus was called on the carpet by the Dean of the College of Science. The Dean began after a perfunctory greeting. "Your previous theories were embarrassing enough, but these latest claims and the manner of their public broadcast, and your unusual behavior on your return, which frankly is schizophrenic and hallucinatory, is totally unacceptable. Speculatus, you are aware, are you not, that you do not have tenure? Your I.Q. may be off the chart but you are impractical and your research is science fiction. This publicity forces me to recommend termination. You will, of course, be granted a hearing before your peers, which is your right under administrative protocol," the Dean said authoritatively. "Until then, continue to teach all your courses," the Dean said, intending that Speculatus should not benefit in the least.

At the word "peer", a wry smile appeared on Speculatus' face. He visualized his peers as he considered in detail what he would say to them. He imagined their verdict. He pictured Dean Murphy as chairman of his hearing. Murphy wasn't the Dean's real name. Speculatus had for years amused himself by substituting fanciful names for the real names of his adversaries and the Dean reminded him of Murphy's Law. Also, the man never had possessed an original thought in his life, Speculatus mused, and to himself he frequently paraphrased the Dean's directives as: "Fellows, there's a harder way to do it, find it".

Suddenly, these thoughts were pushed aside by a brilliant idea. It was ingenious. He would develop a computer Fourier transform[7] of the fossilized sound waves to produce a hologram[8]. That would convince his detractors, he thought with elation! As the mental involvement required to develop the program became great,

Speculatus grew agitated.

"What perverse force drives you?" queried the Dean, observing Speculatus' countenance and assuming his lecture was finally producing the desired effect. "The past rejections of your requests for grant funds and the refusals of your manuscripts should have caused you to seek more conventional approaches."

The Dean's words momentarily invaded Speculatus' consciousness. He forced them out and retreated deeper into unnatural contemplation. The events following his discovery were like a mirror reflecting all his previous life, he thought. He hated the sense of alienation from himself which was brought on by these feelings of failure. It seemed that nothing he did turned out right at the end. His life was filled with near successes, with 'what might have been'. No, that wasn't quite right, he corrected the thought. It was only in the opinion of others that his dreams never seemed to weigh much. The pain of these images drove Speculatus' thoughts inward and to the past. He saw his Sixth grade class, when he had almost won the Lion's Club award for scholarship. He mused and could see his teacher bending down authoritatively, yet almost apologetically, showing him the grade book. That book which, heretofore and hereafter, remained securely in her possession as if its contents were too important to be seen in the original, especially by the students whose marks were recorded there. Speculatus could even visualize on the page next to his name, written neatly in black ink, columns of figures in the high 90s with some 100s. "You had the highest average in class, except for your penmanship. I'm sorry that brought your average down just below Linda's," the teacher said objectively, but needing to explain.

"Get out of my office!" Speculatus was roused to consciousness by the Dean, livid with anger, who was shouting at him, "You haven't listened to a word I've been saying. "It is useless to discuss this further with you. "You are a schizophrenic or you are hallucinating; either way you are hopeless!"

On the appointed day, the Dean brought Speculatus' hearing to order, quieting the embarrassed pleasantries being exchanged among the group, not including Speculatus. A committee of faculty members, accompanied by legal counsel, was seated at a long, polished table in one end of the Dean's spacious office. The Dean read, without coloration, from the folder before him. "The charge against you is unprofessional conduct, Dr. Speculatus. You will be allowed the opportunity to speak later," dictated the Dean, pausing for effect and to pass out copies of several news stories and an official document signed by the college president. "I shall give you all a few minutes to review these," he said. Speculatus sat quietly alert.

The rustle of paper subsided. The Dean looked up and down the table and inquired in his best academese, "Speculatus would you be so good as to tell us briefly about your alleged discovery? We do not need the details of the claims made against you. None of us has been spared the embarrassing news accounts, copies of which I have provided everyone here today, as if such were needed," the Dean intoned. "We wish to know from your own mouth, if you actually said these things, and if so, do you believe them?" At this moment the intercom buzzed repeatedly at short intervals. Immediately, a strange look came across Speculatus' face. The Dean, plainly annoyed at the interruption, rose and crossed to his desk as the "buzz" continued intermittently but incessantly. Speculatus suddenly stood up and gazed piercingly in the intercom's direction. In unison, the eyes of the committee members swung to Speculatus, then back to the Dean who snapped into the intercom receiver, "I cannot now be disturbed". Abruptly, Speculatus cleared his throat and all eyes swung back to him.

Speculatus now had a distant, almost trance-like look in his eyes, and he began abruptly, "When I first obtained the positive reading from the rock face, I knew I had found that for which I had long searched. My sensitive equipment revealed the presence of an unusually ancient sound, although it could not be recorded from outside the rock wall. The sound had been trapped and perfectly preserved because the cave was lined with an unusual magmatic formation which I call 'acoustisprooftus'[9]," he continued. "Of course, I didn't actually detect the sound itself, since none escaped. It is rather technical but in simplified terms, it amounts to an audio-induction *via* magnetic field effects of an imprint of the sound. It was all very weak and without clear harmonics," he explained. "When I later got inside the cave with my special microphones, however, it was a very different story," said Speculatus. "Once I penetrated the acoustisprooftus with my laser, I knew I would have only one opportunity to capture the voice print. It had to be recorded on its first pass. Otherwise it would be immediately lost on breaching the insulating rock formation," spoke Speculatus with considerable agitation.

Several committee members, who had begun smirking, averted their eyes and fidgeted with their folders as Speculatus paused and swept the room with his penetrating gaze. Speculatus continued, "Due to the unique rock formation within the cave, the sound had been fossilized. Perfectly smooth and perfectly vertical walls at each end of the cave had collimated and quantified the sound waves. Consequently, the digitized sound equivalent was reflected back and forth in perfect harmony for 29 million years. Perfectly fitting my hypothesis, an infinitesimal acoustical change in the signal's pitch and volume occurred with each reflection. This is precisely predicted and described by an equation for a new law of Nature which I now propose to call Speculatus' First Law," Speculatus intoned.

"Knowing the speed of sound and the distance between the vertical walls, measured to a fraction of an angstrom, the First Law permits me to date fossilized sounds," Speculatus managed to say the boastful words with modesty. "I have calculated the time elapsed since the utterance of the voice I recorded," he continued.

Here Speculatus turned to the chalkboard and scribbled a formula reflecting his last statements. He drew an equal sign and in large figures wrote 29 ± 0.01 million years. "As some of you may be aware," he began again, "this is almost ten times the age of the oldest previously known human fossil remains. The sound I recorded had persisted in the cave just as light and radio signals persist in space, forever," Speculatus lectured.

"Really, I must object," a loud voice interjected. Speculatus' body jerked visibly. Then he turned from the chalkboard, surprise registering in his eyes. He blinked and turned to see that it was the astronomer Saganian[10] (Speculatus' invented name for this adversary), who spoke. "I won't sit here and listen to this drivel. Your mixture of pomp, calculations and absurd speculations about fossilized sound can not be passed off on me as science, Sir," with exaggerated emphasis on the last word.

Speculatus seemed surprised at the chalk in his hand and embarrassed to be standing. He sat down abruptly without speaking. The astronomer, momentarily distracted by this unexpected and unusual behavior, continued "I long have been a part and, indeed, I am currently a major part of an eminent group of well-funded scientists who are sending coded messages from earth throughout the universe using the 'water hole' frequencies from 18 to 21 centimeters, rather than the 21 centimeter hydrogen emission frequency itself[11]. We have great expectations that some alien culture will receive them. And we are operating, at considerable expense and effort I might add, an antenna at Arecibo in the expectation of listening in on another civilization from deep space. By their very nature, such signals would have to be ancient; but you, Speculatus, are dealing with entirely different phenomena. Sound waves in a cave and radio waves in space are different in billions and billions of ways. Sure, we know that radio waves go on forever in space and we have reason to believe that the humpbacked whale's song may persist around the world undersea to signal a mate, at least it travels further than any other biologically-generated sound, but this cave thing is preposterous," Saganian summated.

The Dean strongly desired to bring the focus back upon Speculatus. "Please, you may continue, Speculatus," he urged. Speculatus still seemed confused and disconcerted. At that moment, the astronomer's wrist alarm went off with a buzzing sound. As he fumbled to silence the alarm, all eyes were momentarily riveted on Speculatus whose appearance suddenly changed again. As if reading from a mental script, Speculatus immediately began where he had ceased before the interruption. "The First Law also has permitted me to re-modulate the signals and to recreate them as they originally sounded. It was an unbelievably tense moment in my lab when at last I was ready to hear the audio from the sound that I had recorded from inside the cave. After computer amplification and re-modulation, all according to my law for fossilized sound, I finally was ready to replay the signal. Hearing that distinctly human voice, frozen in time for so many eons, was the greatest scientific thrill of my life, and I was astonished to hear a coloratura soprano's voice singing Wagner's 'Ride of the Valkyries'[12]," Speculatus related with great feeling.

At this point there was a gasp that ended in a chocking sound. This emanated from the faculty member who had been selected for this hearing by the department of music, as demanded by the college's newly-instituted diversity rules and regulations. Diversity was defined by committee and formed the central theme of political correctness. Speculatus turned toward the sounds, and recognizing the professor, he spoke specifically to him: "Professor Julliard[13] (Speculatus' name for this adversary), I had been fully prepared to find, at most, some grunts and perhaps some half-human, half-ape-like sounds. The finding of this exquisitely beautiful singing voice, however, verifies that human evolution progressed much earlier and more rapidly than any evolutionist has speculated. There is no doubt about it, one must either throw out Leakey's skull 1470[14], or one must throw out this find (Fig. **5**). In fact, all of our current scheme of human evolution must be rethought. "Evolution must be set back in time by a factor of tenfold," Speculatus challenged.

Fig. 5. Leakey's skull KNM-ER1470, next to skull KNM-ER 1813.

"I have done further experimentation," Speculatus continued. "It is still preliminary, but I have made precise measurements of the emanation point of the fossilized sound. From this, I estimated the angle of projection of the voice and the height of the voice box. Thus, the ancient singer was probably only about four feet ten inches tall. Furthermore, she was singing unnaturally, for a human, in a slumped, round-shouldered position. In order further to interpret this data, I have taken the trouble to record a contemporary coloratura soprano in various postures and have established conclusively that only a bent-kneed, knuckle-walking posture is compatible with the particular voice quality of the fossil," said Speculatus.

"Furthermore, analysis of the cranial capacities of contemporary opera sopranos has revealed that opera singing of the quality of the fossil would require a brain case capacity of at least 1125 cubic centimeters, and probably much higher. Thus, we have the emerging picture of a fully human brain and voice, but a very primitive, apelike posture and gait. It is inconceivable that this does not represent the transition to earliest human kind," Speculatus emphasized.

At this point, Julliard was joined by the Nobel laureate biological sciences member of the committee with spluttery sounds best described as pre-explosion noises. Undeterred, Speculatus continued, "I thought it strange at first, myself, to find opera singing at this stage of evolution. But let me play the voice for you." Here Speculatus produced an ordinary appearing pocket tape player and switched it on. Before anyone could respond, there emanated a powerful, yet delicate rendition of the "Ride of the Valkyries" which, nevertheless, possessed a strange and primitive quality.

"Opera in its modern form," Speculatus continued, "is simply a recapitulation of stages of human evolution. I call this my 'Biogenic Law of Operatic Evolution', patterned after Haeckel's biogenic law that 'Ontogeny Recapitulates Phylogeny'[15]. For the non-scientists here, I will say that Haeckel's law simply announces that more highly advanced forms, including humans, pass through all the earlier stages of evolution in the womb, such as human gill slits and tails, he intoned. "Similarly, all operas as we go back in time, can be categorized as less-developed in thematic qualities and less-developed in musical techniques. As we go forward, we find better-developed operas with more complexity. Thus, a scale from Wagner to Puccini—with several dozens of composers in between—logically and scientifically can be established. My fossilized opera lies at the extreme ancient end of my scale of operatic development," Speculatus intoned. Professor Julliard could no longer contain himself. "I am horrified at having opera so represented and particularly by having Wagner so positioned!" he exclaimed.

Speculatus was undeterred; indeed, he seemed not to hear his colleague. "I have analyzed the humidity in the cave and have ascertained that the singer was most probably in the shower at the time of the aria production," he said. "Since more people sing in the shower in the morning than in the evening, it is likely that the fossil was produced in the early morning hours of that long-ago day. Furthermore, analysis of the high notes, in comparison with modern operatic virtuosos, indicates a high probability that the singer was auburn haired. Thus, I have nicknamed her 'Bubbles'[16], said Speculatus, "and I have given her the scientific name *Eopithecus operaii*, which means dawn ape who sings opera, or simply 'Dawn Singer', Speculatus said with a fatherly look in his eyes.

The faculty member representing the music department explosively interrupted here, "Natural selection could not possibly have resulted in creation of this voice at this early stage of evolution." Speculatus appeared physically shocked at the sudden voice sound. "I challenge you to postulate a single, conceivable advantage to opera singing for primitive man – or woman," Julliard said, straining to be politically correct. Speculatus' gaze was focused on the speaker, but he appeared not to follow the musician's meaning. "Therefore, either the voice print is a hoax or it is thoroughly modern in origin. Perhaps an opera singer visited the cave some time in the nineteenth or twentieth century," helpfully offered Julliard, who then paused for effect. The committee guffawed and Speculatus looked embarrassed.

At this moment, the hallway buzzer loudly sounded the end of the last class period of the day. Speculatus' demeanor immediately reverted and he continued, "I have concluded there really is no conflict with natural selection and evolution," he said evenly. "It turns out there is a type of natural selection which I postulate we call 'Speculatus' Third Law: Natural Selection by Deferred Gratification.'"

Speculatus continued, "For example, if a new feature arose which has great future potential but no current value, it would be a big help to the theory of evolution if that feature could be conserved until it is needed. That is, if gratification be deferred. This would save having to produce each evolutionary change at the precise moment it is needed. We should have seen this a long time ago. It is exactly what is needed here and in several other places during evolution," said Speculatus. "This includes evolution of the eye from a faintly light sensitive spot, through many complex, but useless, intermediate stages, to a focusing, image-resolving, brain-connected miracle. Or as a biochemist might cite, the even more inexplicable evolution by complete chance of hundreds of metabolic cycles, many of which must be complete to be anything less than harmful," he lectured. Then, returning to his immediate topic, he said, "after all, opera singing in the shower is hardly a lethal mutation. Once it happened, and anything can happen once merely by chance, it just got preserved. An immediate evolutionary advantage was not required," Speculatus went on to say. "In fact, it is somewhat difficult to find an evolutionary advantage for opera singing at any stage of human evolution, especially, in classical Darwinian terms equivalent to the 'red in tooth and claw' phraseology, borrowed in part by the poet Tennyson[17]," said Speculatus.

Unexpectedly, and to the consternation of all who were present, Speculatus announced that he wished to illustrate his Third Law with a poem he had written. He began to recite his creation:

Gratification Deferred

The simplest imaginable living cell
Might yet spring from the most complex state,
Not quite yet a living cell.
Yet at every stage of advance
Up the ladder of biological complexity,
Is faced the unconquerable enemy:
Irreducible complexity.
If only an Editor could select and choose!
What is to be saved and what to trash,
Then the wasted labor of Hoary Time untold
Might be saved!
Or if gratification be deferred,
And each useless bit,
Discovered first by Blind Chance,
Could only be saved against that day,
However distant,
And however many useless bits required,
When all would fit and work together
To make the whole.

A committee member, silent to this point, pushed back the papers before him and made body movements preparatory to speech. All eyes turned on him. "Well, Ira,", he said, "I'm sure that, as well as everyone in this room, you know my position as the world's foremost astronomer. However, unlike most of my associates, I do not believe in the 'Big Bang' theory. I also favor a steady-state Universe, and that life didn't originate on earth."

At the utterance of his given name, Speculatus blinked rapidly at the speaker. He strained to think but only succeeded in becoming aware that he was trying to think, an odd and shocking awareness. He could not connect up with his last thought. However, before him was the visage of Sir Fred Antibangus[18] who's voice intruded, "There just isn't enough time between the age of the earliest fossils and man, even in the old scheme. Why, if mankind is ten times older, that leaves even less time. This supports my belief that life originated somewhere else in the Universe and drifted here," said Antibangus.

Speculatus, it now appeared to the committee, actually seemed to hear these words which bordered on acceptance of his theory. "Mind you, I'm not saying Speculatus is right about this fossilized opera, just that it fits in with my ideas, about the inevitable consequences of the timing of early evolutionary events, that is all," Antibangus hastened to explain.

The Dean, alarmed at this direction in the proceedings, abruptly said, "Speculatus has made no defense. Do we not all agree?" He glanced nervously around the room to nods of support, and appeared relieved. "Unless you have something else to say ..." He turned and gestured toward Speculatus who showed great agitation.

"I do have a great deal to say", Speculatus said with renewed strength. "I've developed a theory that will allow me to perfect a Fourier transformation of the data which I believe will permit a reconstruction of the sound waves into light waves. If the committee will be good enough to follow me to my laboratory, with a little work, I believe we can see a hologram projection of the ancient singer, herself," Speculatus continued.

There were startled and disturbed looks around the table. The Dean didn't like the way this was going. His mind functioned slowly at best, and under pressure, it sometimes seemed not to function at all. This was one of those non-functioning times. The Dean simply sat speechless.

Dr. Izumov[19], a biochemist who had for years spent most of his time writing science fiction, came to the Dean's rescue. He had remained silent until now, but not from a deficit of wit. He spoke calmly, "Really Speculatus, our patience is at an end. My built-in doubter, about which I have written, has been working overtime. I personally would not walk down the hall to see your hologram, based on the preposterous suggestions you have just made. We have been more than indulgent," he said with finality.

"Quite so, quite so", said the Dean, recovering speech and gaining control. "Speculatus, my field of specialization, before becoming Dean, was Psychiatry. I have more than sufficient evidence to diagnose you as psychotic[20], and I now do so. I am sympathetic to a point and recognize your deteriorating health status and the unfortunate scorpion attacks you have suffered. However, you have made no defense. You are excused, Dr. Speculatus. You will be informed of our conclusions, he said dismissively." Speculatus rose shakily and left the room.

Informed of their conclusions, indeed, Speculatus scoffed to himself as he walked dejectedly down the now darkened hallway. Their foregone conclusions, that is, his thoughts continued. They call themselves scientists, yet they won't even examine my evidence. They will never believe. The evidence they do hear, they discount. The evidence which might convince, they won't even see.

Speculatus opened the door to his laboratory and entered. It grew much later and the science building now was dark. Speculatus sat dejected at his desk in his laboratory. His thoughts spiraled inward. He would lose his appointment and be terminated. He would be terminated from his laboratory and would lose his equipment and instruments. No one else would hire him. He had no wife, no family. There seemed no way out this time.

The loss of sleep and the stress of the past weeks and the lack of food all that day combined to produce a great physical tiredness. His mental depression coupled with this tiredness and dragged him deeper into despair. He was unaware of the passage of time and the hour grew quite late.

A flash of lightening and an almost instantaneous clap of thunder jarred into his consciousness. At that moment, the regional storm warning siren on the campus grounds began to wail just as a violent gust of wind struck against the southwest wall of his laboratory. The weather siren rose and fell in pitch and volume as it rotated to distribute the alarm over a wide campus area. Speculatus seemed at first to be roused by the tumult; then he grew strangely pensive with the repeated rising and falling wail of the nearby storm siren.

Then, slowly, Speculatus came to the realization that a decision had occurred somewhere deep inside. He could not give up. The famous words of Churchill began a drumbeat: never, never, never, give in[21]. A dream that only he shared was better than no dream. With this came inspiration! He could actually generate a computer Fourier transform of the fossilized sound waves and, thus, create a virtual hologram! The thought of actually "seeing" the Dawn Singer drove him onward. Working through the night he completed the mathematical transform and also designed and built a complex instrument that was unique in every respect. Without a moment's hesitation, Speculatus pressed the switch, turning on the instrument.

A holographic image of the Dawn Singer appeared, magically suspended in space. She was three-dimensional, human and entirely lifelike in every respect. Speculatus instantly realized he had been completely wrong about her height and posture; she was exquisite. He turned on the audio, and sound and image merged, as all his faculties were filled as never before with the sensation of being accepted, and Speculatus stepped into the light and took the hand of the Dawn Singer.

SCORPION VENOMS

Before describing the action of neurotoxic scorpion venoms it is helpful to describe the normal physiological functioning of the site in nerve impulse transmission that is affected by these venoms (Fig. **6**). An action potential (nerve impulse) moves down a presynaptic neuron and arrives at a terminus where it can communicate to a muscle (for example). The transmission of the impulse along a nerve is electrical in nature and, for this example, let us stipulate that the "signal" will normally result in muscle contraction. The synaptic cleft is where actions of interest takes place. For the "signal" to get to the muscle cells it must cross a gap, and this is effected by chemical means – a neurotransmitter is involved. The axon terminus of the presynaptic nerve contains vesicles filled with neurotransmitter which must be released and physically move across the synaptic gap. Ca^{2+} enters through open Ca^{2+} voltage-gated channels (the impulse voltage opens them). It is the entry of Ca^{2+}, in our example, that causes the release of neurotransmitter from the vesicles where it has been in storage for that function. Neurotransmitter molecules diffuse across the synaptic gap (which the electrical signal itself cannot cross) and binds to ligand-gated ion channels on the post synaptic membrane. On the postsynaptic side, equally complex reactions occur. Neurotransmitter is bound to ligand-gated ion channels which causes them to open resulting in a graded response in terms of nerve potential (Fig. **7**). The signal is terminated by reuptake and degradation of the neurotransmitter molecules. It can be appreciated that this complex process offers many possibilities for interrupting (blocking) nerve transmission signals to muscles, or for exaggerating or prolonging their actions causing varying degrees of flaccid paralysis, or tetany.

Scorpion venom is produced in specialized glands connected to a means of delivery in the scorpion's tail. Scorpion venoms contain a complex mixture of toxins. When these poisons are injected into a human, for example, a complex interference of nerve transmission results. Voltage-gaited sodium channels (there are many of different specific types) are essential to the production of what are called "action potentials" of nerves and they are the main target of the lethal scorpion toxins. Voltage-gated sodium channels of one (site 1) type bind toxins that interfere with ion transport by physically occluding the pore. These toxins include tetrodotoxin from the puffer fish, saxitoxin from dinoflagellates, and μ-conotoxins from cone snails [3] as described elsewhere in this book. A different sort of action is produced by toxins that bind to site 3. Scorpion α-toxins bind here as do polypeptide toxins from sea anemones and some spider toxins (Δ-altracotoxins) which also are describe in this book. This binding is dependent on membrane potential. There is also a class of scorpion toxins that shift the membrane potential dependence of channel activation by binding to another site. A black widow spider venom component (Δ- palytoxin) acts similarly. These toxins prolong action potential kinetics when bound to a specific site [3].

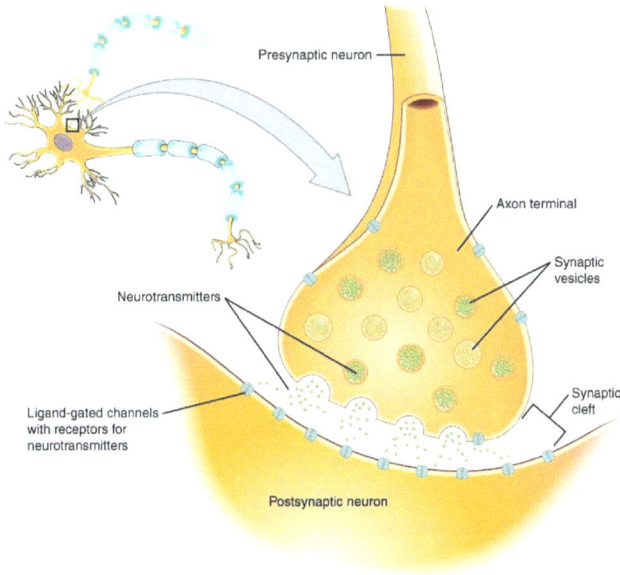

Fig. (6). Mechanisms involved in the 'normal' transmission of nerve impulses.

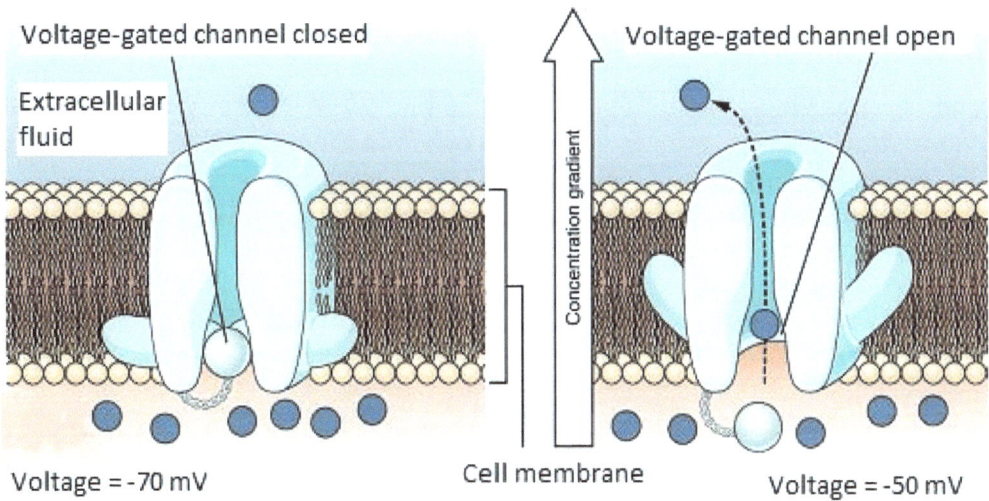

Fig. (7). Voltage differences associated with nerve impulses cause "voltage-gated" channels to open and to close.

There are almost 1500 different known species of scorpions and 50 species are dangerous to man [3]. Scorpion sting can cause medically-important effects ranging from a severe local skin reaction to neurologic, respiratory, and cardiovascular impairments that can be lethal for humans. This site [3] states: "Almost all of these lethal scorpions, except the *Hemiscorpius* species, belong to the family of the *Buthidae*." Of course this includes the death stalker scorpion, *Leiurus quinquestriatus,* and the fat-tailed scorpion, *Androctonus crassicauda*. The death stalker is usually described as "one of the deadliest scorpions in the world" with a sting that is both painful and capable of causing paralysis [4]. This venom is both defensive and offensive. Its toxicity for humans appears to be incidental. The venom assists the scorpion in obtaining food. Crickets are a main food source for this scorpion, and they are quickly subdued by this venom. The toxin in the scorpion venom blocks certain aspects of nerve impulse transmission. An important nerve toxin in the venom of this scorpion acts at the interface of nerves and muscles where nerve impulses stimulate contraction and relaxation.

The movement of chloride ions is integral to this process. These ions move into and out of muscle cells through channels made of specialized proteins that have unique shapes that can control what enters and leaves cells – chloride channels control chloride, specifically. This scorpion venom is also a protein (a very small protein called chlorotoxin which is a chain of only 36 amino acids). The venom also has a particular shape and it fits and binds to certain proteins in the channel opening. It closes these pores in the channels and stops chloride from entering muscle cells. The result is rampant contraction which persists (without relaxation) and the muscle movement is stopped (the muscle is paralyzed). The detail of design is amazing. The small chain of amino acids that make up scorpion chlorotoxin is capable of folding into 256 different three-dimensional shapes but only one works correctly to block the chloride channel. Research on this venom is on-going with potential medical applications for treating brain cancer and ways to safely paralyze patients for certain types of surgery [4]. The relationship of the "grasshopper' mouse (which also likes scorpions), scorpion venom, and pain is complex and interesting [5]. This mouse has an amazing resistance to the stings of bark scorpions which have powerful toxins that readily kill other rodents of similar size to the grasshopper mouse. Knowledge of this relationship and study of this venom is assisting broad research to develop more effective and less addictive pain medications. Two proteins, called nociceptors, are present in normal nerves. When a scorpion stings an ordinary mouse, the venom 'activates" one of these nociceptor protein sites and triggers intense pain along the nerve fibers. In the grasshopper mouse, scorpion venom activities both nociceptors and the pain is blocked. Thus, immunity from pain after a scorpion sting was found lie in the intricacies of these pain receptors and specifically in only one amino acid difference in the proteins [5]. As more is learned about how pain is sensed, there is hope that pain blockers without side-effects can be devised.

Scorpions biosynthesize a variety of toxins; the most toxic species have more than one venom; and some have unusual means designed for the best use of their venom. An unusual example of venoms is found in the Southern African scorpion, *Parabuthus transvaalicus* (Transvaal thick-tailed scorpion, spitting scorpion). It is able to conserve and judiciously use its venom [6]. This site provides references to scientific publications. Making venom is metabolically costly to the scorpion; conserving it is helpful. The poison gland of this scorpion contains a liquid that is mostly potassium chloride (pre-venom) – not particularly toxic. However, when it enters a wound it stings painfully. Imagine rubbing salt into a wound; that is what occurs. Potassium chloride is similar to sodium chloride (table salt). Mice like to eat scorpions and the initial sting of the scorpion injects mostly pre-venom which is painful enough to fend off the mouse in many cases. If the encounter continues, the scorpion will change its venom to a lethal cocktail of poisons. This is an example of the complex behavioral and biosynthetic capabilities of scorpions. Indeed, scorpion venoms are highly complex and most are a mixture of enzymes, peptides, nucleotides, lipids, mucoproteins, biogenic amines and unidentified chemicals [3]. "The bioactivity of the neurotoxins present in scorpion venoms also exhibits a high level of specificity: the venom from a single species of scorpion may contain a toxin preferentially targeting mostly invertebrates, others only vertebrates" [3].

NOTES

[1] The Bible, KJV, Revelation 9:10.

[2] The Doppler Effect is named for Christian Doppler (1803-1853) the Austrian physicist who proposed and described how light and sound waves are affected by the relative motions of a source and a detector. In effect, the wavelength of a sound emitted from a receding object is "stretched" (increased) and thus appears lower in frequency, and the reverse occurs for an approaching object; the changes are proportional to the speeds. Thus, the sound is perceived to change pitch [7].

[3] The lethal dose of the venom for 50% of mice is 5 mg as reported by M. Gwee, S. Nirthanan H. Khoo and L. Chea. Autonomic Effects of Some Scorpion Venoms and Toxins. Clin Exp Pharmacol Physiol. 2002;29:795-801.

[4] The genus name roughly translates to "man killer".

[5] Adrenergic refers to the sympathetic nervous system; more specifically to that portion of the nervous system that functions automatically (controls body functions like heart rate) *via* epinephrine and epinephrine-like chemicals.

[6] Antivenom (the use of the older term, antivenin, is currently discouraged) is prepared by injecting horses with increasing, but initially small doses of venom. It contains proteins (antibodies) that are best kept refrigerated. These antibodies need to be infused over a period of time into a vein to have their neutralizing effect on the venom which spreads throughout the body and attacks the junctions where nerves meet muscles. Scorpion antivenom was first approved by the U.S. FDA in 2011 for a product produced in Mexico [8].

[7] J. P. Hornak has provided a clear description and interpretation of the Fourier transform which is a complex mathematical procedure. It can be compared to a musician with perfect pitch hearing a note and being able to sense what note is being played. A paraphrase of his explanation is: like seeing notes (frequencies) on a sheet of music and understanding them as tones (time domain signals) [9].

[8] The HoloLens® [created by Microsoft] lets you create holograms, objects made of light and sound that appear in the world around you, just as if they are real objects. Holograms respond to your gaze, gestures and voice commands, and can interact with real world surfaces around you. With holograms, you can create digital objects that are part of your world." [10].

[9] An imaginary word that combines "acoustics" and "proof" (in the sense of certain) to indicate absolute retention of sound.

[10] A play on the name of the famous scientist and television personality Carl Sagan.

[11] The so-called 21 cm radio line of hydrogen was predicted by H.D. van de Hulst in 1944, based on quantum mechanics. Whenever a hydrogen atom makes a transition from the higher to the lower energy state, it emits a radio wave at the 21 cm wavelength [11]. Hydrogen is the most abundant element in the universe. Hydroxyl (OH) emits at about 18 centimeters. Hydrogen and OH can form water; thus, the name for the gap between them. Aliens, or ourselves, wishing to communicate would be smart to choose a signal where things are relatively quiet in the universe. The idea that radio signals from earth are reaching distant planets

(and thus, perhaps have been intercepted by aliens) is speculative. It is true that the opening ceremony of the 1936 Olympic Games (the Olympics sponsored by Hitler's Germany) was the first major signal broadcast at a frequency high enough to penetrate the ionosphere of the Earth, and thus to escape to possibly be picked up by alien worlds [12].

[12] This refers to the popular term for the beginning of the third act of Die Walküre by Richard Wagner.

[13] A play on words with name of music professor similar to the famous Julliard School of Music in New York City.

[14] A reference to a fossilized skull (KNM-ER 1470) found in northern Kenya in 1972. It was pieced together from hundreds of bone fragments. It is one of the most controversial human artifacts ever discovered. It was initially dated to be nearly 3 million years old yet it had a large brain and other features unlike any hominid known to exist at the time. The skull was so troublesome that Richard Leaky, the leader of the group who discovered the skull, has been quoted as saying: "Either we toss out this skull or we toss out our theories of early man. It simply fits no models of human beginnings." Leakey later revised the age of KNM-ER 1470 to 1.9 million years, but even then, some scientists have argued that the skull's features are much more humanlike than its contemporary, *Homo habilis* [13].

[15] Haekel's so-called law has long been discredited and was based on fraudulent drawings and interpretations.

[16] The nick-name for Beverly Sills, famous operatic soprano known for her performances in coloratura soprano roles.

[17] These words come from Alfred Lord Tennyson's *In Memoriam A. H. H.*, 1850, Canto 56: "Who trusted God was love indeed - And love Creation's final law - Tho' Nature, red in tooth and claw - With ravine, shriek'd against his creeed." The A. H. H. refers to his friend Arthur Henry Hallam [14].

[18] This is a reference to Sir Fred Hoyle who coined the term the "Big Bang" (a theory of the formation of the Universe) which he, however, coined in derision (which, however, he later denied it was derisive); nevertheless, he did not believe

in the theory, preferring to believe that life on earth originated elsewhere in the Cosmos.

[19] A play of words on the name of Dr. Isaac Asimov, a former professor of biochemistry at a prominent University, who later became a prolific writer and was acclaimed for having written more than 500 books.

[20] A psychotic is a person with psychosis which is a mental illness that greatly interferes with an individual's ability to cope with everyday life, including thought disorder that involves inability to separate fantasy from reality. Symptoms include hearing, seeing and thinking about things that are not physically present. There can be paranoia and delusional thoughts which may be constant or intermittent. Psychosis is known to occur from brain injury or disease and may co-exist with schizophrenia or bipolar disorders; and it is reported to result from certain chemicals by drug use [15].

[21] On October 29, 1941, U.K. Prime Minister Winston Churchill visited Harrow School to hear the traditional songs he had sung there as a youth and was invited to speak to the current students. Churchill's speech included these words: "You cannot tell from appearances how things will go. Sometimes imagination makes things out far worse than they are; yet without imagination not much can be done… this is the lesson: never give in, never give in, never, never, never in nothing, great or small, large or petty – never give in except to convictions of honor and good sense [16]".

REFERENCES

[1]   Quintero-Hernández V, Jiménez-Vargas JM, Gurrola GB, Valdivia HHF, Possani LD. Scorpion venom components that affect ion-channels function. [cited 2017 Mar 17]; Available from: https://www.ncbi.nlm.nih.gov/pmc/articles/PMC4089097/pdf/nihms578515.pdf

[2]   Linaker MR. Scorpion: Second generation. , New English Library 1982; p. [cited 2017 Mar 19];158 p. Internet Available from:http://eol.org/pages/3195189/details

[3]   Bosmans F, Tytgat J. Voltage-gated sodium channel modulation by scorpion alpha-toxins. Toxicon [Internet] NIH Public Access , 2007 Feb; [cited 2017 Nov 23];49(2): 142-58. Available from: http://www.ncbi.nlm.nih.gov/pubmed/17087986

[4]   Turnbough M, Martos M. Venom. ASU- Ask a Biologist [Internet]. [cited 2017 Nov 23]. Available from: http://askabiologist.asu.edu/venom/scorpion_venom

[5]   Collins F. Gain Without Pain: New Clues for Analgesic Design | NIH Director's Blog [Internet]. [cited 2017 Nov 23]. Available from: https://directorsblog.nih.gov/2013/11/07/gain-without-pain-new-c-ues-for-analgesic-design/#more-2246

[6]   A Toxic Tale: This Scorpion Can Make Two Kinds of Venom | Mental Floss [Internet]. [cited 2017 Nov 23]. Available from: http://mentalfloss.com/article/63467/toxic- tale-scorpion-can-make-two-kinds-venom

[7]   Doppler effect | physics | Britannica.com [Internet]. [cited 2017 Mar 20]. Available from: https://www.britannica.com/science/Doppler-effect

[8]   FDA Approves First-Ever Antivenom for Scorpion Stings [Internet]. [cited 2017 Mar 20]. Available from: http://www.medscape.com/viewarticle/747536

[9]   Hornak J. CHAPTER-5, The Basics of MRI, Fourier Transorms [Internet]. [cited 2017 Mar 18]. Available from: http://www.cis.rit.edu/htbooks/mri/chap-5/chap-5.htm

[10]   https://developer.microsoft.com/en-us/windows/holographic/hologram

[11]   Spectroscopy of the 21cm Line [Internet]. [cited 2017 Mar 22]. Available from: http://astro.u-strasbg.fr/~koppen/Haystack/spectro.html

[12]   How far into space do our TV signals go? [Internet]. [cited 2017 Mar 22]. Available from: https://briankoberlein.com/2015/02/19/e-t-phone-home/

[13]   Controversial Human Ancestor Gets Major Facelift [Internet]. [cited 2017 Mar 22]. Available from: http://www.livescience.com/7224-controversial-human-ancestor-major-facelift.html

[14]   "Red in tooth and claw" - the meaning and origin of this phrase [Internet]. [cited 2017 Mar 19]. Available from: http://www.phrases.org.uk/meanings/red-in-tooth-and-claw.html

[15]   Medical Definition of Psychosis [Internet]. [cited 2017 Mar 18]. Available from: http://www.medicinenet.com/script/main/art.asp?articlekey=5110

[16]   Never Give In [Internet]. [cited 2017 Mar 18]. Available from: http://www.winstonchurchill.org/resources/speeches/1941-1945-war-leader/never-give-in

CHAPTER 4

Poisoned by Lovely Plants

Abstract: Many plants contain chemical compounds that are mildly toxic to humans; some plants are overtly poisonous, and a few are deadly. Throughout history, the leaves, roots, stems, and berries of certain plants have been used for murder, often of the vilest sort. Also, people and animals are accidentally poisoned by plants some of which are garden-variety ornamentals. Because there are so many choices, in this chapter, I have elected to describe certain plant toxins because of extreme toxicity, some because of unusual examples of their murderous applications, and a few because they have become the subject of legends. I hope to draw the reader's interest in the science of toxic plants and their poisons and about the use of these poisons in modern accounts and in tales that are mostly myth. A true story, but one with sensationalized nuances and uncertainties and a modern revisiting, is the murder conviction of Harvey Crippen based on a death from a medicinal chemical derived from the belladonna plant. I will explore what I have chosen to call the art and science of five very poisonous plants: belladonna (deadly nightshade), white snakeroot, castor bean, rosary pea, and monkshood. I will include references rather than extensive descriptions of the plants, details about the signs and symptoms of poisoning, an example of poisonous use (extensive for belladonna), and the biochemistry and biology of the mechanisms of toxicity of the chemicals.

"Lady Nancy Astor: Winston, if you were my husband, I'd poison your tea. Churchill: Nancy, if I were your husband, I'd drink it."[1]

Keywords: Alkaloids, Belladonna, Belle Crippen, Castor bean, Hawley Crippen, Deadly nightshade, DNA analysis, Ethel Le Neve, Georgi Markov, Hyoscine, Hyoscyamine, Jequirity, Monkshood, Plant toxins, Scopolamine, Ricin, Rosary pea, Vitali test, Water hemlock, White snakeroot, Wolf's bane.

INTRODUCTION

Reliable information about poisonous plants in the United States can be found in a report assembled by the Division of Botany of the U.S. Department of Agriculture [1]. In this chapter, I will give an overview of specific toxins, selected plants, and historical and forensic cases of poisoning by plant toxins.

As described in Chapter 1, there are many ways of assessing poisonous substances. I believe there are good reasons to consider how little of a poison is

Olen R. Brown

required to kill an adult, otherwise healthy, human; how long it takes to kill; and whether there is an easy, reliable diagnosis with a sure antidote or other effective treatment. I have proposed that instead of assessing poisons by weight of the lethal dose, it would be better to determine the lethal dose based on the number of molecules of a substance required for lethality.

Plants, depicted as the seven most deadly plants, are described in the on-line Encyclopedia Britannica: water hemlock (*Cicuta maculate*), deadly nightshade (*Atropa belladonna*), white snakeroot (*Agerativa altissima*), castor bean (*Ricinus communis*), rosary pea (Abrus precatorius), oleander (*Nerium oleander*), and tobacco (*Nicotiana tabacum*) [2]. Plants have been found to synthesize many chemicals (perhaps 100,000 or more) some with pleasing, medicinal, or poisonous qualities including the well-known caffeine, theobromine (found in chocolate, tea leaves and the kola nut), cocaine, and morphine [3]. Let us examine several of these poisonous plants and their toxic chemicals including their intriguing histories, in more detail.

BELLADONNA, THE DEADLY NIGHTSHADE

This chapter begins with a murder story often described as the second most notorious in British legal history (behind Jack the Ripper). The tale involves scopolamine (also called hyoscine), derived from belladonna, the deadly nightshade (Fig. **1**). I recall that as a small boy in southwestern Oklahoma I was enticed by the purple flowers and small berries of a plant that I sometimes played with−a local variety of the deadly nightshade. I had been warned by my mother that it was poisonous but one day I punctured a small berry and it squirted liquid and seeds into my eye. I was frightened about consequences but only suffered stinging pain and an eye effect that was similar to what I later experienced at the optometrist office when my pupils were dilated. I since have learned that this plant was not the deadly nightshade but perhaps one of a diverse group of plants consisting of more than 2,500 herbs, shrubs, and trees that include belladonna. Some of these plants are now "naturalized" in North America [4]. The plant I experienced was a small weed with purple flowers containing yellow centers and yellow berries; it most probably was not significantly poisonous.

Atropa belladonna, native to Eurasia, is a perennial with greenish berries which ripen to a lustrous, deep black. Other names for the plant are: belladonna (Italian for beautiful lady), deadly nightshade, devil's berries, death cherries, beautiful death, and devil's herb [5]. It contains the alkaloids scopolamine (hyoscine) and hyoscyamine. Alkaloids are chemical substances that contain basic (in the chemical sense of not acidic) nitrogen atoms with non-clear boundaries between them and other similar substances. Alkaloids are produced by many organisms including bacteria, fungi, plants and animals and many have medicinal uses or toxic properties. This chemical property will be a confusing element in the chemical identification of the poison in the Crippen murder case which will be described.

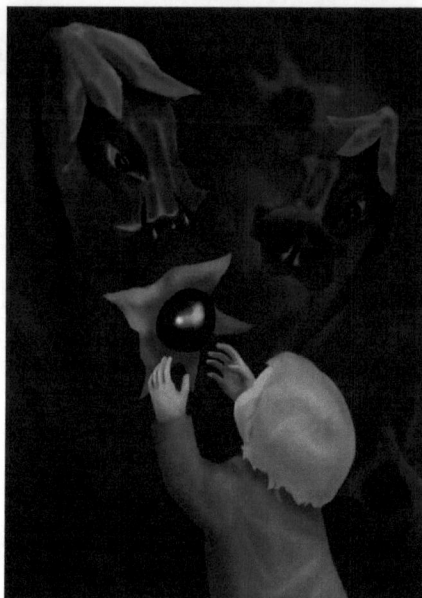

Fig. (1). The deadly nightshade, *Atropa belladonna*. Original art by Claire Engler.

Many accounts call belladonna 'the stealthy poison of assassins' because a person can develop tolerance from taking small doses. This would set the stage for a stealthy assassin to take a sip of a drink in the presence of an intended victim but a drink by the victim would be deadly. In the original story of Banquo and Macbeth (better known version in Shakespeare's play), this pair plotted to kill King Duncan after winning his favor by helping in a battle fought and won against the Danes. The story in Cliff Notes for Macbeth [6] relates that they slaughtered the entire Danish army by cunningly mixing a sleeping potion administered in drinks supplied to the army; once asleep they were readily killed while helpless from the effects of the drug. Supposedly from this, Shakespeare got his idea of having Lady Macbeth give a sleeping potion to the guards of King Duncan's chamber. That the sleeping potion contained belladonna may be a stretch, but it seems a likely candidate.

A SENSATIONAL MURDER STORY MORE THAN 100 YEARS AGO

Hyoscine (scopolamine, obtained from the belladonna plant) was the poison of choice in a story aired in 2008 on PBS: *"Executed in Error, Hawley Crippen"* [7].

The story was about a sensational murder than occurred in 1910. Even the young Winston Churchill, then Britain's Home Secretary, was said to have been avidly following the story. A modern, forensic toxicologist and key investigator who revisited the case stated: "The Crippen case was the O.J. Simpson case of 1910. I don't think any murder case in history had been covered that much in the newspapers." Indeed, at the time, the story was being sensationalized and read world-wide. Another expert for the PBS airing expressed a troubling doubt about the original jury verdict that stemmed from a central element of the case evidence – the victim's body was dismembered. This expert had never known of a poisoning case with such a grizzly follow-up. His opinion was that poisoners typically tried to make the death look like an accident [7]. In the reexamination of the murder, court records were examined and tissue from a histopathology slide of the victim, still available after nearly a hundred years, was examined using modern forensic techniques including DNA analysis. A team of experts in poisons, genealogy, and DNA analysis traveled between the United States and England, where the murder occurred, to thoroughly reevaluate the case. Lacking a source of the victim's DNA, they compared DNA from the histopathology slide, with DNA from certain female relatives located by genealogy searches. The DNA 'didn't match' and more startlingly it was from a male, and the victim, Crippen's wife Cora (also known as Belle Elmore), was of course a female [7]. Is it ironic that the victim's name was Belle and the suspected poison was scopolamine which is present in the belladonna plant?

The PBS forensic team focused on the new forensic finding and pieced together conclusions from police and court archives that included what they described as 'suppressed' documents [7]. This included a letter from Belle to Crippen claiming she was living in America and had no plans to appear in public or otherwise to save him from execution. This letter was deemed a hoax at the time of the trial; however, there is no evidence it was shown to Crippen's lawyers. Based on these findings, James Patrick Crippen, the closest living male relative of Crippen was said to be "… formally requesting that the British government pardon the doctor and return his bones to America." Before his execution, Crippen had penned what the PBS program called 'eerily prophetic words' in a letter to a friend: "Face to face with God, I believe that facts will be forthcoming to prove my innocence" [7].

Let us return to those days of yesteryear and see what light can be shed on this controversy, and on belladonna poisoning. In 1910 the wife of a homeopathic physician, Dr. Harvey Crippen (Fig. **2**), disappeared. Dr. Crippen was having an affair and when the object of his affection moved into his home, questions were raised about the disappearance of his wife, Belle. Crippen was investigated by Scotland Yard, found guilty, and hanged. During the trial it was determined that he killed Belle with the plant alkaloid hyoscine which is synonymous with scopolamine [8]. It is an alkaloid found in the leaves and seeds of several plants including *Atropia belladonna*. It exerts anti-cholinergic actions similar to atropine but with a greater CNS effect. It also is found in henbane (*Hyoscyamus niger)* as well as some other plants in the *Solanaceae* family [9]. The beautiful but deadly nightshade, also called belladonna, and the lowly potato belong in this plant family.

Fig. (2). Dr. Hawley Harvey Crippen.

Sources consistently report most of the case facts I present here but interpretations vary [7, 10 - 13]. An interesting novel, based on the murder has been written by John Boyne [14] and there is an on-line description of the book [15]. I have drawn freely from all these sources; however, my personal slant and interpretation occasionally may be obvious. None of the authors of the cited reports should be blamed if I get it wrong. It seems obvious that no one will ever know exactly what happened or which side of the controversy is correct. In the book: The Trial of Hawley Harvey Crippen [11], the Introduction states: "The region of human morality is not a flat plain; there are hills and valleys in it, deep levels and high levels... somewhere between these extremes, far below the highest, but far above the lowest, lies the case of Dr. Crippen, who killed his wife in order to give his life to the woman he loved." This is but the opinion of one man, Filson Young. However, in my view, Young's opinion carries weight because he lived in this period and edited the trial notes to include his own comments. I consider that the facts presented in Young's account [11] are as accurate as can be found and I will insert comments in the ensuing descriptions where I recognize they differ from other, including more recent, writings about the case.

Some background is helpful in understanding the events of the murder. Hawley Crippen was born in 1862 near the beginning of the Civil War. Except as perspective, this date is of no relevance. His parents were Myron Augustus Crippen and Andresse Skinner Crippen, which is relevant. They were prosperous owners of a dry goods store that provided a comfortable family home-life and they apparently passed on a strict, Protestant work ethic to Hawley. Having been interested in medicine since childhood, Hawley chose to obtain an M. D. in homeopathic medicine. Filson is probably more accurate in identifying this as a diploma from the Hospital College of Cleveland Ohio, as an ear and eye specialist at the ophthalmic hospital. Harvey moved to New York where he met and married a young Irish nurse, Charlotte Bell. It is ironic and odd happenstance that her last name, Bell, is similar to the 'Belle' which his second wife assumed as a given name and that the poison used is found in the belladonna plant. They had one son, Otto, and Charlotte died of apoplexy in early 1892. Filson says this date was 1890 or 1891 [11]. It is said that Hawley coped poorly, and persuaded his parents to care for Otto. Remaining in New York, he met Cora Turner (who chose to go by the name Belle), and they married in the fall of 1892. Filson [11] states: "In 1890 or 91 his wife died and later he made the acquaintance of a girl of 17 known as Cora Turner". Filson also adds the information that "They lived in Philadelphia but Cora stayed in NY to take voice lessons, and pursue her passion to become a grand opera singer."

Cora was described, in the various accounts, as born in Brooklyn to immigrant parents of Baltic descent who named her Kunigunde Mackamotzki which she subsequently changed to Cora Turner (and she later took the name Belle). She left home at 16, pursued her vocal and acting talents and started a theatrical career. This was in the late 1800s and tells us much about the venturesome spirit of Cora. When she and Hawley met, Cora was 19 and an aspiring opera singer under the name Belle Elmore. Most accounts suggest Belle was an unlikely pairing for the Protestant, mild-mannered Hawley. Some suggest that Belle was motivated in the relationship by Harvey's title and potential wealth as an M.D., and he was captivated by her show-business persona.

Hawley's previously promising future in homeopathy was on the wane and Cora (Belle) probably found her expectations of their standard of living to be unfulfilled. They may even have lacked money for her to continue dramatic lessons and the nationwide economic depression closed theaters and Belle's dreams of stardom evaporated. Hawley's failing medical practice resulted in him finding employment with a homeopathic mail-order business. By hard work and ingenuity, Harvey succeeded. By 1895 he was general manager in Philadelphia and in 1897 he opened the company's over-sea's office in London. This date is given as 1900 by Filson [11]. This move is relevant to the life and events that unfolded. Belle remained in Philadelphia and according to most accounts she had numerous, amorous affairs. She subsequently moved to London and pursued her stage career, although by some accounts her talents were scant, and the couple's marriage apparently was stressed. For reasons that are not completely clear, but which may have been related to their marriage difficulties, some of which related to Belle's career pursuits, Hawley was fired from his job. He was unsuccessful in finding adequate employment and this increased tension between them when they were forced to move from fashionable Piccadilly to low-rent Bloomsbury.

Thing turned much worse when Hawley 'met' the then 18 year-old Ethel Le Neve who was a nurse at the homeopathic clinic where Harvey worked. It is said that Ethyl was the opposite of Belle; she was a demure, intelligent 'English rose'. Hawley, apparently, was enamored and Ethel was taken by what was, apparently, Hawley's mild, industrious manner. It is said that by 1903 they were inseparable and in spite of Hawley's Protestant convictions, their relationship was described to have progressed to physical intimacy. However, I did not find a reference to this in Filson Young's book [11].

Hawley's economic status was improving and he and Belle moved in 1905 to a home in the more fashionable Hilltop Crescent in Holloway and Belle 'starred' in the Music Hall Ladies' Guild and her 'American ways' made her a successful fundraiser for the Guild. By all accounts, Hawley and Bell by this time were living 'entirely separate' lives – Hawley with his passion for Ethyl and Belle with alleged affairs and theater friends.

Supposedly for extra income, but perhaps with an ulterior motive, Belle insisted they take in lodgers and this ended badly in late 1906 when Hawley discovered Belle in a compromised position with one of the student boarders. I also did not find this in Filson's account. It is said that this resulted in Hawley and Ethyl consummating their relationship, which had apparently been platonic until this point. Subsequently, although Hawley and Belle continued to live together, the marriage was described as entirely broken. It is said that Belle was unsure of the identity of Hawley's lover, Ethyl, until the time when he left homeopathy to pursue dentistry and took Ethyl with him as secretary. A further shock came when Belle learned that Ethyl was pregnant. Hawley was said to be delighted and broached the subject of divorce with Belle; however, Ethyl miscarried. It is further said that Belle decided to play the virtuous, betrayed wife. By 1909, living together at Hilltop Crescent was nearly intolerable for Hawley and Belle. There were frequent arguments and Belle threatened to ruin Hawley's career and reputation. It is alleged that Hawley decided he could stand it no longer and he began a murderous plan that may have been initiated when Hawley told a close colleague, Dr. John Burroughs, that he had concerns about Bell's health.

Records show that on January 17, 1910, Hawley ordered five grains of hydro-bromide of hyoscine from a chemist who supplied medicines for the dentistry practice which he was now following. Hyoscine is another name for scopolamine.

It was used in those days in very small dosages for various purposes including subduing mental patients; however, there were no standard uses in dentistry. Five grains is approximately 325 mg, about the weight of one aspirin tablet.

The preceding 'triangular' relationship and the immediate events surrounding Belle's death were accepted as strong evidence against Hawley by the jury which convicted him. However, Hawley insisted he was innocent and a recent reinvestigation raised doubts about the circumstances of Belle's death and this will be described subsequently.

On the day after the last sighting of Belle, as established by trial records, Hawley came to his dental office and attended his patients as usual. He told Ethyl that Belle had left him; presented Ethyl with a gift of Belle's jewelry; and pawned the rest of her jewelry. Hawley arranged for Ethyl to deliver a note from Belle of resignation as treasurer from the Ladies' Guild. He provided the excuse (reason) that Belle had to go to America to nurse an ill relative. This excuse was said to have created suspicion which grew outspoken when Ethyl, wearing some of Belle's jewelry, attended a Guild Ball accompanied by Hawley.

Hawley tried to stop the gossip by saying that Belle had fallen seriously ill in California. He sent a telegram to this effect to friends (the Martinettis) with whom Harvey and Belle had shared a final meal before Belle's "disappearance" (more about this later). Then, Hawley and Ethyl left the scene for a short vacation together in France. This was presented at trial as extremely odd and incriminating. Upon their return, in response to persistent inquires from Belle's friends, Harvey 'made-up stories' including that Belle's body was cremated in the United States. Belle presumably would not have wanted cremation because of her Catholic background. With mounting suspicions, the Ladies' Guild approached Chief Inspector Walter Dew of Scotland Yard who found there was insufficient evidence to bring charges. It is established that the Ladies' Guild, in Agatha Christie style, relentlessly pursued their own investigation. They discovered that no boat had sailed for the United States on the day Belle was alleged to have travelled there, and no one named Crippen had died in California on the day claimed by Hawley. Drew was literally forced to take further action and he questioned Hawley on July 8, 1910. Strangely, it is recorded that Harvey admitted that he had made up the story of her death, claiming that Belle had left him for another man and he was trying to avoid scandal and spare her good name. Apparently, Drew was satisfied. However, Hawley is said to have panicked.

He persuaded Ethyl that they must leave England for a year to let the 'scandal' die down. They went to Antwerp the next day to board a boat for Quebec. Dew, on July 11, discovered their absence and finding Crippen's maid had been dismissed and that the home was being prepared for an extended absence of the owners, Dew organized a thorough search that extended for two days. The rotting remains of a body were discovered beneath the cellar floor. The account published in *History Today* [13] details that:

> "Police... found the gruesome remains of a body beneath the coal cellar. Wrapped in a male pyjama jacket, which was later identified as Crippen's, it had no head, no limbs no bones and no genitals, but there were traces of a poison that Crippen was discovered to have bought not long before Cora's disappearance."

A medical examination of the torso revealed a scar from a healed operation consistent with Cora's (Belle's) medical history. Hyoscine, a poison, was also found forensically. On July 16, 1910, arrest warrants were issued for Hawley Harvey Crippen and Ethel Le Neve.

The intrigue continued as the case was splashed in newspaper headlines, with detailed stories and pictures of the couple, throughout England and parts of Europe. Deciding to travel incognito (which was neither illegal nor that unusual for 'celebrities'), they boarded the SS Montrose in Antwerp for Canada on July 20, travelling as 'Mr. Robinson and son' with Ethyl, to little purpose, crudely disguised as his son. The Captain of the ship, Kendall, had seen a local newspaper with pictures of the pair. The crude disguise attracted his attention and he was suspicious of the odd couple. On July 22, Captain Kendall (Fig. **3**) sent a wireless

Fig. (3). Captain Kendall.

telegram to the White Star Line in Liverpool, indicating that Hawley and Ethyl were aboard his ship. This information was immediately passed on to Inspector Drew at Scotland Yard. This is said to have been the first time that this new means of communication (the wireless telegraph) was used in the apprehension of a criminal (Fig. **4**).

Fig. (4). A photograph of an early example of the wireless telegraph, the communication method used to apprehend Hawley and Ethyl while fleeing aboard ship.

Unfortunately for the Crippens, and usefully to Inspector Dew, another White Star Ocean Liner, the SS Laurentic, was about to leave Liverpool for Quebec, the destination of the Crippens. It was a faster ship and was scheduled to arrive before the SS Montrose. This transatlantic chase was followed by the newspapers which splashed stores of the progress of the two ships along with stories of a love triangle (Fig. **5**). The daily plots of the progress of the two ships showed that the SS Laurentic was steadily overtaking the Montrose, and it actually reached Quebec the day before the SS Montrose. Inspector Dew boarded the SS Montrose, disguised as a harbor pilot, and he arrested Hawley and Ethel using his authority as a Scotland Yard detective. They were in British territorial waters which made this possible. If Hawley had sailed to the United States instead of Canada, as an American citizen, Dew could not have arrested him.

Fig. (5). A newspaper front page from the time, announcing the murder.

The Trial of Crippen and its Modern Reinvestigation

The team of investigators reported (PBS program, 2008) on new findings from reexamination of the case [7, 15]. A key finding was that the DNA obtained from tissue presented at trial as Belle's remains, was not that of Belle. Furthermore it was from a male. It is reasonable to be able to distinguish male from female by DNA analysis, even on a hundred year-old tissue sample. On this point, however, John Boyne, author of a novel about this case [14] in which he invents a character who could have killed Belle, is reported to have stated the opinion [15]:

> "I don't think the DNA evidence has cleared anything up… There were suggestions Crippen could have been an abortionist and the body in the cellar was one which went wrong. He still dressed his mistress as his son and fled. He obviously had something to hide. Dr. Crippen is one of those mysteries that will never be solved."

The above-related discounting of the DNA evidence is one explanation of the reported finding that the DNA was from a male [7].

I will proceed by recounting the trial as reported in the book available on-line: 'The Trial of Hawley Harvey Crippen' (edited, with notes and an introduction by Filson Young, 1920) [11], and the book, with portions available on-line by John Ensley: 'Molecules of Murder: Criminal Molecules and Classic Cases' [12].

First, I will give a brief overview of the trial [10]. Harvey Crippen and Ethyl were tried separately in October of 1910 in London. Harvey refused to allow Ethyl as a defense witness, based on Harvey's apparent concern for her reputation. He was found guilty after only 27 minutes of deliberation by the jury. The presiding judge, Lord Alverstone, sentenced Crippen to death by hanging. The appeal of Crippen's sentence was refused and he was executed at Pentonville Prison in London. An account of Harvey's execution by Richard Cavendish [13] states that he was hanged at 9 am on November 23, 1910. The rapidity with which trial, conviction, and hanging occurred may be surprising in today's world. Cavendish reported [13] intimate details of the last hours of Harvey including that he spent the hour before his hanging with the Roman Catholic prison chaplain and two wardens who had forestalled a plan by Harvey to commit suicide using broken glass from his spectacles. After this Harvey was said to be calm but unable to finish his breakfast. The hangman, John Ellis is said to recall that "Crippen came across to me as a most pleasant fellow". He was said to be smiling when the shroud was paced over his head preparatory to hanging. He died instantly and was buried in the prison graveyard.

Ethyl's trial, beginning shortly after Harvey's, was on charges of accessory to murder after the fact and being a fugitive from justice. It lasted only one day and she was found not guilty after only 12 minutes of deliberation by the jury. It is recorded that Ethyl visited Harvey every day and sent him a letter after each visit. Harvey's estate was bequeathed to her. On the day of Harvey's execution, Ethyl left England for New York. She later moved to Toronto and was employed as a secretary. She subsequently returned to England, married, and passed away in 1967 at age 84. The Crippen home, where the remains were unearthed, remained empty for many years and was destroyed in a German air raid during World War II. A side note is of interest. Kendall, captain of the SS Montrose, escaped death when the ship he then captained, the SS Empress of Ireland, was wrecked in 1914 with loss of more than 1,000 lives. This sinking occurred, ironically, at Father Point in Quebec which is very near the site where Hawley had been arrested only four years earlier by Kendall who was one of only a few survivors of the sinking.

Details of the Crippen Trial

A most interesting and complete record of the trial is available [11] and I reference it to supplement the story related here. As part of the published 'Notable Trials Series', this account is described as: 'Edited, with notes and an introduction by Filson Young' with the publication date 1920. The Crippen case was justifiably included in that series because:

> "In a case of such world-wide notoriety, the theme inevitably of much speculative and imperfectly informed discussion, it is more than ever useful to have the facts, in so far as the trial revealed them, set forth exactly as they were unfolded to the judge and jury … If the trial is less interesting from a legal point of view than some others, this defect is atoned for by the extraordinary human and dramatic interest with which the story is packed, and which has placed Dr. Crippen in the front rank, so to speak, of convicted murderers."

I shall draw primarily from the 'trial book' [11] as my source for the trial itself, believing it to be the most accurate and reliable account available. To more fully document this account I shall sometimes provide page numbers, which will refer to this book and I shall include quotations where the sense might be lost by my interpretation. I shall not follow the trial sequentially because that would require a book-length account which the interested reader can obtain elsewhere (Fig. **6**). My focus is on the forensics of the time and the poison itself.

Crippen's attorney, Mr. Tobin, gave the closing argument for the defense in the trial with criticisms of the prosecution's case including that the remains alleged to be that of Belle Crippen nee Elmore were "… buried in clay which practically excluded all air… and with lime added as well. There was no proof that it was

Belle Elmore, or even of a woman." This was indeed prophetic of future events. He also argued strongly that the 'scar' which I shall describe later, was not a scar at all (p. 148).

Mr. Tobin also refuted that the 'gummy' substance, identified in the visceral remains of the corps alleged to be that of Belle Crippen, was not proven to be hyoscine as was testified to by Dr. H. W. Willcox for the prosecution. Willcox said that this substance was not proven to be a plant alkaloid and if it were, it could have been any of three alkaloids: atropine, hyoscyamin, or hyoscine which he alleged were not chemically distinguishable. [Note: hyoscyamine and hyoscine are alternate spellings used at the time]. This defied a linkage the prosecution had made with a medicinal, hyoscin, that Harvey Crippen had admitted purchasing shortly before Belle's death (p.149). This testimony reveals the precarious state of forensic conclusions based on chemical analysis of the time period.

Fig. (6). A depiction of the trial of Harvey Crippen.

Dr. Reginald C. Wall was examined on the stand by Mr. Tobin for the defense. Wall's credentials included that he had a Master of Arts and Doctor of Medicine from Harvard; that he was a Fellow of the Royal College of Physicians, London; and a member of the Royal College of Surgeons. He professed other accomplishments including that he was one of several pathologists serving the

London Hospital and the author of various medical publications. He gave the firm opinion that the skin tissue described as a scar was not a scar. This is important because if it were a scar it linked the remains to Belle because of a known operation. It is interesting that the trial record indicates that a juror was allowed to ask the expert, Wall, a question (p. 137). The question, itself, is unimportant.

Wall's testimony, as elicited by attorney for the defense Tobin, included the following question about the testimony for the prosecution by Dr. Willcox. "He [Willcox] extracted a gummy substance characteristic of hyoscine and not of hyoscyamin, atropine, or any animal alkaloid?" The questions pointed to the reliability of a specific forensic test (the Vitali test) and led to doubt about whether 'animal or vegetable alkaloids' were proved to be present in the viscera of the remains recovered from Crippens' cellar. Dr. Wald stated that the Vitali's test was relatively new and on the issue (apparently, because I find the testimony lacks clarity) of the certainty of proof that hyoscine was present, he appears to conclude: "… so we do not know; no one knows." Wall continued: "A second test was performed by Dr. Willcox in which… small, round spheres were produced from the chemicals isolated from the viscera".

Whether these were characteristic of hyoscine alone, or to be found with both hyoscyamine and atropine, another expert for the defense, Dr. Blyth said: "I have not been able to get them. I have attempted to get what Dr. Willcox has stated according to the descriptions that have been forwarded to me, but I must confess that I have not been able to distinguish between atropine, hyoscyamin, and hyoscin by hydrochloric acid, as Dr. Willcox has done. No one knows whether these round spheres might be produced at last in the case of animal alkaloids."

I report this testimony here in some detail because I have found no other recounting of the trial testimony that placed doubt on the finding of the poison hyoscine in the remains. This is of paramount importance, in my view. Hyoscine is connected to a purchase of this medicine by Dr. Crippen. Perhaps significantly, Dr. Willcox had been made aware of this purchase, and it guided, or at least informed, his chemical examination and search for poisons in the remains. Indeed, it is known that 'animal alkaloids' (products of bacterial action on animal tissue) exist and whether these could have been the source of the positive Vitali's test was strongly introduced by the defense. This point was either not understood adequately by the jury or, alternatively, it made no significant difference in their minds (p.140). I will return to this later in this chapter.

Hawley Crippen was extensively examined in the dock at trial. He responded that "he did not write out prescriptions." This is consistent with his role as a homoeopathic physician, but he admitted that his 'medicine cabinet' contained:

hyoscine, aconite, gelsemium, and belladonna (p.127). Crippen also stated: "Well, the first I knew of hyoscine as a prescription, as a treatment [seeming to correct himself] was in 1885. I have been treating patients chiefly for ear troubles for a very long time, but not in a general way." He also said he prescribed hyoscine for nervous diseases, coughs of a septic character and for asthmatic complaints. He went on to describe how he used the 5 grains of hyoscine he was proven to have purchased shortly before Belle's disappearance. This explanation included that 1/3 of what he had purchased "… was left in his office". This hyoscine was never found, and his testimony about the poisons apparently was not convincing to the jury as an explanation for why he possessed these poisons, including the one alleged to have killed Belle (p.120).

The testimony of Dr. Willcox was very damaging to Crippens. Willcox described in detail his efforts to identify poison in the tissues he examined. He said this began on July 23 with examination of the stomach, kidney, and liver. It is important to remember that he had been made aware of the purchase of hyoscin by Crippen. This was not unusual, as a toxicologist needs direction to know what to search for chemically. However, it can introduce bias, especially in interpreting the extremely subjective tests used in those days. I do not say that it did. His words, recorded at the trial are the best evidence we have regarding identification of hyoscin, and I will provide them in some detail.

Most of the stories written about the trial do not focus on the reliability of the report that hyoscine was found in the remains. It is important to reflect on the state of chemical forensic analysis in 1910. Willcox states (in part): "I first of all searched for mineral and organic poisons. I found traces of arsenic in the intestines and liver, and I found traces of creosol (carbonic acid) in intestines and liver, small traces." He said he attached no importance to these and they were from disinfectants used. He said he "… commenced examining for alkaloidal poisons" (which required 2 to 3 weeks). He extracted (using usual processes for extracting alkaloids) tissues from weighed portions of stomach, intestines, kidney, and liver. He found alkaloids present in all the extracts. He then applied further tests to "… see what kind of alkaloid was present. I tested for all the common alkaloids: morphia, strychnine, cocaine, and so on… and found that a mydriatic alkaloid was present; that is an alkaloid the solution of which if put in the eye of an animal, causes the pupil to enlarge and dilate." (p. 67).

It may be interesting to note that this procedure (instilling test liquids into the eyes of laboratory animals) was actually part of some tests in those times; people ordinarily experience a similar effect at the ophthalmologist today when their eyes are dilated to better examine the retina.

Dr. Willcox continued to testify about how he tested for poisons. "I applied a further test, and found that it was mydriatic vegetable alkaloid, of which there are 3: atropine, hyoscyamine, and hyoscine. I applied further tests, and found that the alkaloid that I had got in the extracts <u>corresponded</u> to hyoscine [emphasis added]. I have no doubt it was hyoscine. I could tell that in two ways, one by examining the residue with a lens and microscope; it was gummy, there were no crystals there. Another way was by adding to a solution of the residue some bromine solution-hydrobromio acid [the transcript says hydrobromio; I assume this is hydrobromic] and I got round spheres, but no crystals. Hyoscin gives spheres exactly alike I got. Atropine and hyoscyamin give needle-shaped crystals. The two ways I have described—the gummy residue and the spheres from the bromine solution—pointed to hyoscine only. In the stomach there was one-thirtieth of a grain, and in the kidneys there was one-fortieth of a grain, in the intestines, one-seventh of a grain and in the liver one-twelfth of a grain, there was a trace in the lungs. The total amount in all organs submitted to me was two-sevenths of a grain." [A grain in approximately 65 mg; a regular aspirin contains 325 mg of aspirin and a baby aspirin has 81 mg, for comparison]. He continued to say: "Hyoscin is not used medicinally in the form of hyoscin. It is a gummy syrup stuff, which it would be impossible to handle, and so a salt is used. The salt which is used is the hydro bromide of hyoscine, and it is the preparation listed in the British Pharmacopeia. In the whole of the organs submitted to me the amount of the hydrobromide of hyoscine was 2/5 of a grain, which could certainly correspond to more than half a grain in the whole body. What is a fatal dose? From a quarter of a grain to a half a grain."

Half a grain is approximately 33 mg, a significant amount. However, this was calculated based on an undoubtedly large (unreported) factor used to convert detected amounts in small portions of extracted tissues to amounts in the entire body. This obviously introduced a considerable uncertainty in the reported amounts. In today's court (an unfair comparison) quantities could not be reported to the jury without an indication of the uncertainties.

Words by the Lord Chief Justice appear here, including the following. "If a fatal dose were given, it would perhaps produce a little delirium and excitement at first, the pupils of the eyes would be paralysed; the mouth and throat would be dry, and then quickly the patient would become drowsy and unconscious and completely paralyzed, and death would result in a few hours." It was then stated that a proper dose *via* injection was one-hundredth to two-hundredth of a grain. The lord Chief Justice continued to say: "As far as I know, it is not used as a homeopathic remedy. It is rather salt and bitter, and it can be administered in something with a pronounced flavor, such as stout or beer or sweetened tea or coffee, or it could be given in spirits" (p.69). This informs us about the nature of scopolamine for surreptitious use.

Cross examination of Dr. Willcox by Tobin continued and Willcox answered: "I have tested for hyoscin before, but I believe this is the first case where the question of murder has arisen. I have never found hyoscine in extracts of dead bodies before this case... There are two classes of mydriatic alkaloids ... vegetable and animal, which are produced after death by the action of the putrefactive bacteria without any being introduced into the body during life." It is significant to note that the records of the trial show that: "...18 days before he did the test, Willcox had been told that Dr. Crippen had bought some" (p. 69). Dr. Willcox then stated that he did not discover sufficient alkaloid to do a melting point test.

Identification of a substance based on its melting point was the 'gold standard' for proof of the chemical identity of an unknown substance at that time. It was as good as was possible in those days; color test results were more subjective, and less reliable. At best as used in this case, the color test, when negative, could reasonably exclude many substances with almost absolute certainty, but a positive test is only consistent with a specific substance, not proof of its presence.

I will now summarize the scientific (forensic) evidence produced at trial that apparently resulted in the jury believing in Crippen's guilt. Most of the written commentaries I have read focus on other testimony pointing to the guilt of Dr. Crippen; I shall focus primarily on the forensic science and toxicology. It should be noted that any review of trial science from 1910, done in 2017, runs a huge risk of being over-critical.

Vitali's test was primarily used by Dr. Willcox to identify hyoscin. This test, first described as having been proposed in 1880, was described in a paper written in 1933 [16]. This test has subjective elements of interpretation of colors and changing colors, and cannot be definitive, in my opinion, for identifying hyoscine in extracts from decaying tissue remains. However, it was the best that could be done at the time. A relevant publication indicating the state of such analysis was published in 1912 and entitled: "The Destruction of Alkaloids by Emulsions of the Body Tissues" [17]. This paper reports recovery and identification of small amounts of atropine (a plant alkaloid) in the range of 0.1, 0.01, and 0.02 mg, with lower amounts designated only as "present". This suggests a lower limit of reasonable detection in extracts, but because volumes of tissue extracts were not reported it is not possible for me to relate this to the test results reported at the trial. Translated to grains (1 mg is approximately 65 grains) as reported by Dr. Willcox, the smallest amount (0.01 mg) is very small suggesting the technique could assess tiny quantities of this poison.

Forensic Evidence that Convicted Hawley Crippen

Other accounts have focused on the circumstantial evidence and testimony that figured prominently in the conviction of Harvey Crippen. This evidence was substantial and undoubtedly it was sufficient, and perhaps determinative, for the jurors. I will not detail it here but it included the following: (1) the human remains of a murdered victim unearthed beneath the floor of the Crippen home; (2) Crippen's "guilty flight" in crude disguise soon after the body was found in the cellar beneath his home; (3) Crippen was a homeopathic doctor with medical knowledge; (4) records showing purchase by Crippen of the poison hyoscine with some of this poison found in his medicine chest; (5) the motive provided by the triangular relationship between Crippen, his wife Belle, and Ethyl; (6) Crippen told several stories, that were shown or accepted as untrue, to explain the absence of Belle; (7) wide knowledge that the marriage of Crippen and Belle was "unhappy" in many ways; and (8) Belle could not be located.

FOUR TRAGICALLY POISONOUS PLANTS

Introduction

I have selected four plants for very different reasons. White snakeroot is included, although it rarely causes poisonings today, because it historically was a scourge perhaps equally as frightening as cholera or typhoid to early settles in the America of the 1800s. The deadly ailment had no known cause; there was no effective treatment, and certainly no cure. It is historically relevant that the birth-mother of Abraham Lincoln died a painful death from the toxins of this plant transmitted by cow's milk when Abe was only 9 years old.

White Snakeroot (*Argeratina altissima*)

White snakeroot is less toxic than the other plants we describe in this chapter, and knowledge about the toxic chemical is incomplete and somewhat controversial. *Argeratina altissima*, previously known as *Eupatorium rugosum*, is reported in some parts of Indiana and Ohio to have caused up to 50% of the deaths of early settlers in the early 1800s [18]. Livestock are poisoned at doses of 1 to 1.5% of their body weight of plant matter ingested daily for 1 to 3 weeks and the disease is called "trembles" [19]. The history of knowledge about the chemicals involved in poisoning has been recounted in the cited article. Initially the agent was called tremetol and it was believed to be a pure compound with the structure of $C_{18}H_{32}O_3$. Later work showed it was a complex mixture of terpenes, sterol and ketones including the benzofuran ketones tremetone, dehydrotremetone, and other related chemicals. Testing of the toxins in cell cultures grown in the laboratory indicated that the likely toxin was tremetone; however, animal models have not proved this.

Davis, *et al*. in 2016 [19] reported that "The incidence and severity of poisoning was not correlated with total doses of tremetone or total benzofuran ketone concentrations suggesting they are not closely involved in producing toxicity and [pointed to] the possible involvement of an unidentified toxin."

Fig. (7). White snake root in bloom.

White snakeroot is a member of the daisy family native to eastern and central North America, and it is found along wooded streams and in pastures (Fig. 7). Its name may derive from an erroneous belief of early American settlers that the rhizomes (underground root-like structures) were useful in treating snakebites. However, both the rhizomes and leaves are very toxic and human fatalities have occurred from drinking milk from cows that had grazed on the plant [20]. Before its white flowers appear, the plant is inconspicuous.

Nancy Hanks Lincoln, the birth-mother of Abraham Lincoln, was poisoned by the toxin of this plant [21]. Dr. Thomas Barber of Bourbon County, Kentucky was perhaps the first to describe this condition and he noted: "… a man turns sick and his domestic animals tremble" [22]. Anna Pierce Hobbs Bigsby (1808-1869), known affectionately as Doctor Anna by those whom she served, is credited with establishing the cause [23]. Anna took courses to become a physician, a very rare thing for a woman in those days. In 1828 she was the only physician in the county where she grew up, Hardin County, in southern Illinois. She was soon confronted by an epidemic called milk sickness which killed both animals and people including Anna's mother and her sister-in-law. Determined to find the cause, Anna discovered that it was seasonal, beginning in the summer and continuing until after first frost, and it most frequently occurred in cattle. Legend says that a Shawnee American Indian woman (name not recorded) told her that the white snakeroot caused milk sickness. Anna tested this by feeding the plant to a calf and saw the characteristic sickness develop. Although correct in her deduction, she had received no official recognition when she died in 1869. It is said that the medical community did not officially recognize the plant as the cause of the illness until sometime in the 20th Century [23].

The death of Nancy Lincoln poignantly describes one death, typical of many [20, 24, 25]. The Lincolns lived on Pigeon Creek in Kentucky; the time was 1818. Two neighbors had died of similar illnesses within the prior week. It was known, even then, as milk sickness although the cause was not; it was also called puking fever, swamp sickness, and river sickness. No one knew its cause, it was devastatingly painful and debilitating, and there was no cure. Various causes were proposed over the years and several plants were suspected before it was proven to be snakeroot. The specific culprit includes a chemical called tremetol, a long-chain alcohol containing 16 carbons (ordinary 'drinking' alcohol has 2 carbons). The signs and symptoms of snakeroot poisoning in humans progresses to Cheyne-Stokes respiration with breaths becoming progressively shorter and more shallow and with longer intervals between breaths. Blood pressure and temperature drop and respiratory death occurs. Prior to death the stomach, intestines, kidneys and liver are highly inflamed. Excruciating pain and symptoms continue for 10 days or more and there are changes in the heart tissues and atelectasis in the lungs impairs oxygen transport leading to death. Death by this poison is made most gruesome and frightening because of the symptoms, their progression, the unknown cause, and the lack of any successful treatment.

It is said that Nancy Lincoln had instilled in young Abe the virtues of compassion and honesty and that she was the source of Abe's intellectual curiosity [26]. Although Nancy had little or no formal education she stressed the importance of reading and learning in the young Abe Lincoln who was only 9 years-old when his mother died. We can only imagine the effect of her horrible and unexpected death on Abe who later called her affectionately his 'angel mother'.

Caster Bean (*Ricinus communis*)

Poisoning by the caster bean plant is said to be rare today. However, its specific toxin, ricin, is potent and lethal. It is generally considered to be among the most toxic, naturally-occurring poisons. It has been used as a weapon of mass destruction. These factors are reasons to elevate castor bean on the list of plant poisons. The beans are also used to make castor oil, which itself does not contain the toxin [27]. The beans are highly toxic when consumed but are frequently used ornamentally in prayer and rosary beads and in musical shakers (maracas). The plant (Fig. **8**) originally was known in Africa and Asia but has spread to most temperate and subtropical regions including the southwestern United States where it occurs along stream banks [28]. The ease of access to this poison is said to have made its use for suicide unfortunately popular. The U.S. Communicable Disease

Center lists ricin as a Category B terrorism agent. In 1980, it has been stated, based on considerable evidence, that the Iraqi government made weapons-grade ricin and tested it on animals with delivery by artillery shells. It was used in a highly-publicized terrorism incident in which it was delivered to the office of Senator Bill Frist [28].

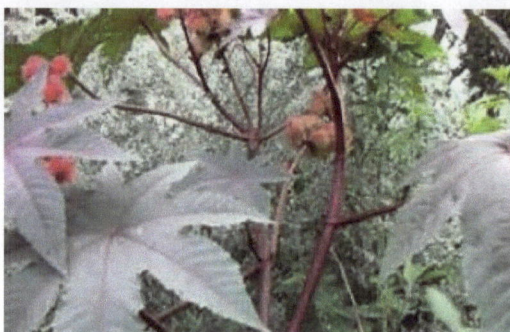

Fig. (8). The beautiful but deadly castor bean plant.

The bean's poisonous chemicals are toxalbumins that inhibit protein synthesis. Severe toxicity occurs to multiple organ systems. The specific toxin, ricin, is present throughout the plant but concentrated in the leaves. Ricin produces severe gastroenteritis, followed by delirium, seizures, coma, and death. The beans also can cause allergic skin rashes. Poisonings, however, are said to be rare in the U.S. and the Poison Control Center recorded no deaths in 168 cases. Deaths, however, have been reported to occur from as little as one bean ingested by a child [28]. Poisoning symptoms, by the ingestion route, are delayed with a latency of about 3 days but symptoms may continue for 10 days with nausea, vomiting, diarrhea, and abdominal cramps. There may be liver toxicity and renal failure. Drowsiness and seizures have also been reported. When eye contamination occurs, blindness can result. Inhalation of dust containing the toxin can cause serious illness within 8 hours [28].

Ricin Mechanism of Toxicity in the Body

Much is known about the site and mechanism in the body where ricin exerts its deadly effects. It has been studied because of its curiously intense toxic activity and because it is a strong candidate, with known uses, as an agent for chemical warfare, terrorism, and political threat and assassination.

Ricin is a large protein (Fig. 9) and its depiction as a ribbon diagram emphasizes how a long chain of amino acids forms a complex, three-dimensional structure. This size and complexity allows it to bind very specifically to protein components in structures called ribosomes that are 'factories' for synthesizing proteins using information ultimately provided by DNA (acting as a blueprint or template), but using copies made in RNA. More specifically, the ricin B chain binds the toxin to sites on the surface of human cells. The A-chain of ricin inactivates ribosomes by cleaving a glycosidic bond on the large subunit of the ribosome at residue 4324 in a region called the sarcin-ricin loop.

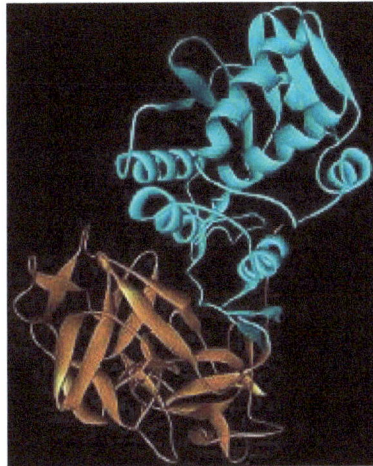

Fig. (9). Ribbon diagram showing three dimensional protein structure of ricin with the A and B chains shown in blue and orange, respectively.

At the molecular level within cells, ricin stops protein synthesis which is lethal to cells (Fig. **10**). This mechanism also occurs with abrin and will be described additionally in connection with abrin poison later in this chapter.

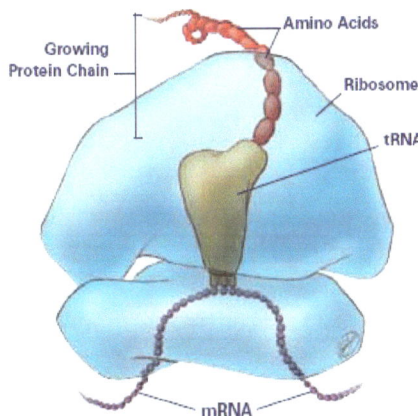

Fig. (10). Ricin poisons protein biosynthesis by binding to a specific site on the large subunit of the eukaryotic ribosome.

Thus, ricin and abrin share this important mechanism of cellular toxicity which has been investigated in detail [29]. The clinical manifestation of toxin exposure is severe lung inflammation and respiratory insufficiency. These authors found that the magnitude of depurination by abrin was lower than for ricin.

The Strange Case of Georgi Markov and Ricin Poisoning

This mysterious occurrence involving assignation, cold war intrigue, and James Bond type gadgets, has been told by PBS [30] and by Kelly Hignett, a lecturer at Leeds Beckett University [31]. I have drawn heavily on these accounts.

On the morning of September 7, 1978, Georgi Markov while waiting for a bus on his way to work at the BBC was struck in the thigh by what was later determined to be a small pellet. He described the pain as "… similar to an insect bite." A man, described as heavyset with a foreign accent picked up an umbrella from the ground, mumbled: "I'm sorry" and hurried away. Markov, who had a small, painful bump on the back of his thigh became progressively more ill during the day and was admitted to hospital that evening with a high fever. He died on September 11, the fifth day after the incident.

Based on his signs and symptoms, an autopsy, and much deductive work (and some assumptions), investigators concluded that Markov was assassinated in "… an operation conducted by the Bulgarian Secret Service (the *Darzhavna Sigurnost*) with guidance by the Soviet KGB" [31].

Georgi Markov was described as an author, broadcaster and communist-era dissident [31]. He was born in 1929 in Sophia, studied chemistry in the 1940s, worked as a chemical engineer and school teacher but became an acclaimed novelist, playwright and TV script writer. Some of his writings were critical of communistic Bulgaria and were banned by the regime in power in Bulgaria. In 1968 Marko left Bulgaria and settled in London where he learned English and became a broadcast journalist for BBC and for Radio Free Europe and published acclaimed novels and plays. He defected to the West and became unwelcome in Bulgaria in the extreme and was sentenced in absentia to six years in prison. He continued to criticize the regime *via* the BBC and from 1975 to 1978 he worked on the 'In Absentia' reports. Todor Zhivkov declared him an enemy of the Bulgarian regime. The *Darzhavna Sigurnost* kept a surveillance file on Markov (code name: Wanderer) and he received more than one telephoned death threat. Markov's publisher later said that "Markov knew his activities made him a possible target for assassination" as reported in Hignett's account [31].

Accounts of the assassination provide similar details [30, 31]. By autopsy an odd 1.52 mm pellet made of platinum and iridium was found in his leg associated with the site of a puncture wound. It is proposed that the pellet was fired by compressed air and not by gunpowder from possibly an umbrella (Fig. **11**). The pellet was determined to be a jeweler's watch-bearing and was about the size of a pin-head (Fig. **12**). Two holes, each said to be 0.34 mm in diameter, had been drilled to create an 'x-shaped' well in the pellet.

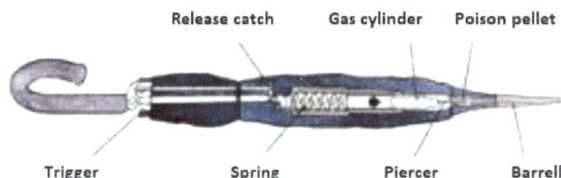

Fig. (11). Proposed umbrella weapon used in Markov assissanation.

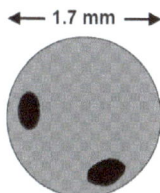

Fig. (12). Drawing of the metal pellet deduced to have contained the Ricin inside a cavity made by connecting small holes.

The pellet was extremely hard metal and the investigators surmised that the holes had to have been drilled with a type of laser by 'spark erosion'. They calculated that the pellet could have contained 1/5 mg (0.2 mg or approximately 13 grains) of substance; and that it probably was sealed with a wax that melted at body temperature to release contents from inside the pellet [30]. However, they could find no trace of ricin, the suspected poison, in the pellet, and they found no ricin in Markov's body. However, intelligence agents had for decades suspected that the Soviet Union had been investigating the use of ricin as a chemical warfare agent and it was said to have been at the top of the list of two or three toxic agents suspected [30].

Hignett [31] provides similar details but she said it was concluded that the "… pellet had been filled with 0.2 mg of the deadly poison Ricin…" Her story provides a photograph of the pellet and an interesting account of the assassination from which I will draw extensively. Hignett recounts that evidence suggests that Markov's assassination was ordered with the full knowledge and involvement of both the Bulgarian *Darzhavna Sigurnost* and the Soviet KGB. Two previous attempts to assassinate him failed: a toxin slipped into his drink at a dinner party and a possible attempt during a visit to Sardinia. September 7 was the birthday of

Todor Zhivkov's (communist head of state of the People's Republic of Bulgaria), and Markov's murder was somehow to be "… a gift to the Bulgarian leader". Hignett further states "… recently declassified Bulgarian Secret Service files confirmed the close nature of the relationship between the *Darzhavna Sigurnost* and the KGB representatives were keen to ensure there was no 'trail' directly linking Markov's death to Moscow." Hignett cites a *Time* magazine story [31] that listed Markov's murder as one of their 'Top 10 Assassination Plots' and in 1998 Bulgarian President Peter Stoyanov described it as 'one of the darkest moments' in communist Bulgaria. *Time* also reports that in June 1992, General Vladimir Todorov, a former Bulgarian intelligence chief, was sent to jail for 16 months for destroying 10 volumes of material on the Markov case and an unnamed suspect committed suicide rather than face trial for destruction of these files.

Curiously this story is similar to the account of Vladimir Kostov, a Bulgarian exile, who approximately 10 days prior to the assault on Markov, was struck by "… the same kind of metal pellet embedded in his skin." while waiting at a Paris metro station. He developed a high fever and was hospitalized but recovered [31].

Rosary Pea (Jequirity) and Abrin

Rosary Pea is also known as jequirity, crab's eyes, cock's eyes, prayer bead, love pea, and by other names (Fig. **13**). It contains abrin which is similar to ricin (Fig. **14**). It is said that a single bean has enough abrin to kill an adult human (by comparison, 7 berries from belladonna are required) [3]. A report from the Florida Center for Aquatic and Invasive Plant Division is informative. The rosary pea occurs in Florida as an ornamental but is native to India and parts of Asia. Its seeds were regarded to be sufficiently uniform that it has been used as a standard of weight. Another source says its seeds are also used in jewelry; but a single seed contains sufficient abrin to kill an adult human [32].

Fig. (13). Jequirity beans (rosary pea); note the colors and "eyes" which are typical; the beans also have a very hard shell and if not crushed or chewed can pass through the digestive system without releasing the toxin from inside the beans.

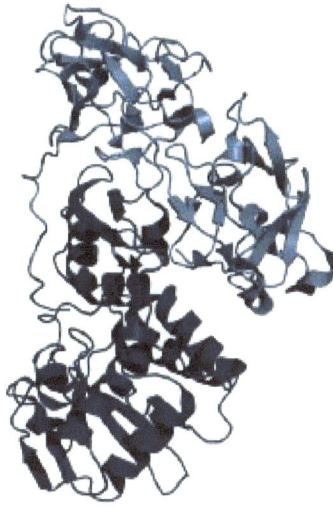

Fig. (14). Ribbon diagram of abrin showing the 3-D structure of the protein molecule.

Mechanism of Toxicity of Abrin

Both abrin and ricin share the mechanism of toxicity previously detailed for ricin: the cleavage of specific chemical parts of subcellular structures (site-specific depurination of ribosomes) to stop proteins synthesis (Fig. **10**). This is lethal to cells.

An article by Kumar and Karande in 2016 [33] describes a recombinant vaccine that has been developed and which protects mice from abrin lethality. It is said to be a promising immunotherapeutic. An earlier paper published in 2015, described the A and B protein chains of abrin toxin and reported a recombinant antibody to the B chain subunit. This antibody induced "… a good immune response after 4 immunizations," suggesting that immunization against this part of the toxin "… may be a promising" as a vaccine candidate for humans. Also in 2015, a paper was published showing that antibody treatment against pulmonary toxicity from abrin confers significantly higher levels of protection than treatment against ricin intoxication [34]. This paper states: "To date, there is no established therapeutic countermeasure against abrin intoxication." It further states that: "Due to its high availability and ease of preparation, abrin is considered a biological threat, especially in context of bioterror warfare." This paper also describes intoxication by ricin and abrin to be similar and to include: "… massive recruitment of neutrophils into the lungs, high levels of pro-inflammatory markers in the bronchoalveolar lavage fluid, and damage to the alveolar-capillary barrier." In contrast, the protective effect of anti-abrin antibody treatment was found to differ significantly from that of anti-ricin treatment. While anti-ricin treatment efficiency was quite limited, anti-abrin treatment conferred high-level protection at 24 hour post-exposure and polyclonal anti-abrin antibodies were effective even as late as 72 hours post-exposure. The differential ability of the anti-toxin treatments to dampen inflammation caused by the two similar toxins, abrin and ricin, were said to explain the radically different levels of protection achieved following antibody treatment.

A study reported in 2015 [35] indicated that the abrin A-chain is the major factor triggering the response that leads to a process called apoptosis which results in cell death. The B-chain is required for internalization of the toxin into targeted cells. Their results were also interpreted to mean that inhibition of protein synthesis is the major event contributing to abrin-mediated apoptosis.

Abrin has other toxicity in addition to removing specific adenine residue from the 28s rRNA of the 60s-ribosomal subunit as was stated less specifically previously and as was shown in Fig **10**. Abrin also has demyelination effects on the brain and this was found to result, at least in part, from induced oxidative damage and increased reactive oxygen species, glutathione depletion, and increased lipid peroxidation [36]. There also were changes in neurotransmitter concentrations in brain and cortical brain regions showed demyelination after abrin exposure. The authors concluded that abrin poisoning leads to neurodegeneration and neurotoxicity through oxidative stress, inhibition of acetylcholinesterase, lipid peroxidation, and decreased amounts of myelin (a protein in nerve cells).

A case report of abrin poisoning appears to be typical [36]. Clinical management by primarily supportive measures with administration of fluids, anti-emetics, and activated charcoal (depending on the time since exposure) was recommended. It was stated that there is no antidote for abrin poisoning. In the case reported, a 22-month old child had ingested approximately 20 rosary peas from a 'peace' bracelet. The child's primary signs were episodes of forceful emesis. She was tachycardic with apparently what was considered normal blood pressure, elevated blood urea nitrogen and serum creatinine. The child was said to have been "… resuscitated with normal saline IV and received anti-retching medication" and was well at discharge 24 hours later. This case (reported in 2014) was said to have been the first case of human abrin toxin poisoning confirmed by the quantification of L-abrin as a biomarker. L-abrin designates a specific chiral form that is biologically active.

Abrin is reported to be highly toxic, with a human fatal dose of. 0.1 to 1 microgram per kilogram of body weight (this calculates to 7 to 70 micrograms for an average person [37], and these authors found that highly purified abrin by the intraperitoneal route had a LD_{50} of only 0.91 micrograms. Various blood substances were altered including WBC and RBC counts. Liver peroxidation was increased and there were degenerative changes in several organs.

Monkshood

Monkshood contains the poisonous aconitine, a neurotoxin whose cellular and molecular site of action is at membrane sodium channels essential for nerve impulse transmission [3]. The flowers of this plant are strikingly beautiful but remind some of the hooded clothing worn by medieval monks (Fig. **15**).

Fig. (15). Monkshood.

It is considered to be the most poisonous plant in Europe and it has been called the 'Queen Mother of Poisons' [38]. The plant has also been called wolf's bane and devil's helmet. It has been alleged that in WWII Nazi Germany coated bullets with aconitum derived from Monkshood [3]. Historically, Giovanni Battista Porta [1535-1615] in 1589 wrote the treatise: *Neopoliani Magioe Naturalis* which described poisoning by various plant substances including aconite, as cited by Jolle Steele [9].

While conducting a biological inventory survey of public lands near the Current River in Shannon County, Missouri, a new genus of wildflower, monkshood, was discovered by botanists [39]. The article says monkshood is native to Georgia, Indiana, Kentucky, Maryland, North Carolina, Pennsylvania, South Carolina, Tennessee, Virginia, and West Virginia but does not mention its toxicity.

A curious account in 2014 [40] reports that: "A gardener collapsed and died after apparently handling a highly-poisonous plant… a coroner has heard". This is an ambiguous way to sensationalize the story which also states that the coroner and a histopathologist concluded that "…it was more likely than not that the [victim] died after having come into contact with the deadly purple flowering plant". Against normal procedures, the initial blood samples taken at admission of the victim to the hospital were destroyed. The story (with no references given) states: "The attractive plant has also been responsible for several human deaths, including the Canadian actor Andre Noble who died on a camping trip in 2004 after consuming the plant." The article also relates: "In 2009 Brit Lakhvir Singh, dubbed the 'Curry Killer', poisoned her lover… with a curry dish laced with Indian aconite, from the same plant."

An on-line article entitled simply 'Monkshood' gives interesting facts about this plant [41]. It notes that European monkshood (*Acontium lycoctonum*) was once known as wolf's bane and was used to poison both wolves and criminals, and supposedly was a component of witches' brew. Words of the English romantic poet John Keats [1795-1821]: provide a serious warning:

> "No, no! go not to Lethe, neither twist Wolf's-bane, tight-rooted, for its poisonous wine; Nor suffer thy pale forehead to be kist By nightshade, ruby grape of Proserpine; Make not your rosary of yew-berries…"[2]

Wolf's-bane is monkshood, and there is reference to nightshade and to rosary beads made of yew-berries (not of jequirity beans), all in a melancholy setting.

Several accounts allude to monkshood in Greek myths. It was said to have sprung from the slobber of Cerberus when pulled from the cave of the underworld by Hercules. Because of this association with this hound of hell it is sometimes called dog's bane (or wolf's bane) (Fig. **16**). Athena in legend used its poison to turn *Arachne* into a spider [42, 43]; see Chapter 7. Ovid apparently told the story that the slobber from the mouth of Cerberus was an ingredient in the poisons of the Erinyes and of the sorceress Medea [44].

Fig. (16). Cerberus, from Greek Mythology, whose slobber was said to have created the first aconite.

The Strange Case of George Lamson

George Henry Lamson was convicted in 1881 of murdering his brother-in-law, Percy John, reportedly the first homicide using aconitine [41]. The story of the trial was published in 1913 in a book edited by Hargrave Adam [45]. The book is part of the series 'Notable English Trials' to which I previously have referred. I have also drawn from the account [38] 'M is for Monkshood' and Kathryn Harkup's book: 'A is for Arsenic: The Poisons of Agatha Christie' [46]. Harkup's book has entertaining stories about 14 poisons that appear in Agatha Christie's own books and includes monkshood. It is said [46]:

> "Agatha Christie used poison to kill her characters more often than any other crime fiction writer ...The poison was a central part of the novel, and her choice of deadly substances was far from random; the chemical and physiological characteristics of each poison provide vital clues to the discovery of the murdered. Christie demonstrated her extensive chemical knowledge (much of it gleaned by working in a pharmacy during both world wars) in many of her novels, but this is rarely appreciated by the reader."

A curious method was reportedly used to test for poisons at autopsy. Dr. Thomas Stevenson (1838-1908) was an expert in alkaloid poisons. He was used as an expert for a case in which a man named Percy John allegedly was murdered. Stevenson examined the victim's remains. He made extracts of tissues and since there were no chemical tests to identify aconitine (which he suspected was the poison administered to kill Percy) he used an unusual skill he possessed. It is said that Dr. Stevenson had an extensive knowledge of the taste of perhaps 50 to 80 alkaloids which he had sampled and had the amazing ability to recall and thus identify. It is also said that, as a party trick, he was able to identify by taste a particular alkaloid before a colleague could identify it using chemical tests. This astounding ability was widely known and established his reputation. He claimed that the 'taste' of aconitine was unique (to him) and that as little as 1/60 of a grain (approximately 1 mg) could be fatal [38]. It is similarly reported [41] that: "One of the interesting facts of the case is the methods used for analysis of the poisons. One method was taste: they applied some of the alkaloid obtained from the body to their tongues, which produced a biting and numbing effect; a precisely similar effect was produced by a similar application of aconitine."

Lamson, who was convicted of the crime, was a medical doctor who had established a practice in Bournemouth, England. Unfortunately, he had acquired a morphine habit that was said possibly to have resulted during an earlier period of service as an army surgeon in Romania and Serbia. This led to a declining medical practice and he acquired considerable debts [38]. An inheritance brought some financial relief which, however, was short-lived. It is said that he hatched a murderous plan to obtain a second inheritance by killing Percy John an 18-yea--old relative who was paralyzed from the waist down, but otherwise in good health.

Dr. Lamson apparently made an unsuccessful first attempt to murder Percy in 1881. While on holiday on the Isle of Wight, Lamson gave Percy a pill. Shortly afterward Percy was violently ill but made a full recovery and returned to his boarding school for the fall term. It is said that Lamson's money worries became acute and while in America that year he purchased a special type of gelatin capsule designed for administering powdered medicines. On November 24, 1881, Lamson bought two grains of aconitine (approximately 130 mg) from a London pharmacist who made no record of the purchase which was improper for dispensing a poison.

On December 3 of the same year, Dr. Lamson visited Percy at his boarding school in Wimbledon. Sherry was served and Lamson added a spoonful of sugar with an odd claim that it reduced the effects of alcohol. Dr. Lamson also served three slices from a 'Dundee' cake he produced that was already sliced, taking the last slice for himself [38]. Lamson showed some of the capsules he had obtained in America, and recommended that the Headmaster use them as a means of giving bitter medicines to his pupils. He then is said to have demonstrated its use by filling one of the capsules with sugar, from the same bowl he used for the sherry. He gave the capsule to Percy and asked him to show the headmaster how easy it was to swallow these pills. Percy did as he was instructed and Dr. Lampson left abruptly, making the excuse that he did not want to miss the train to his boat to France. Within a few minutes after Lampson left, Percy became quite ill; he had stomach pains, vomited and was carried upstairs to his room. Percy is reported to have said that he felt the illness he had experienced when Dr. Lamson had given him a pill as previously described. He became more ill and convulsed; was attended by two physicians but died later that night. The doctors believed he was poisoned by some vegetable alkaloid and they were suspicious that Dr. Lampson was the culprit. The police were summoned but Lamson escaped to France. He later returned to England, voluntarily surrendered and was arrested for murder [38].

At trial [45], the autopsy of Percy did not produce a cause of death. The pharmacist who sold aconitine to Dr. Lamson testified. Despite keeping no record of the sale, he said the sale was so unusual that it 'stuck in his mind' and when he read about the case in the newspapers he contacted the police. Dr. Lamson's personal notebook also recorded symptoms he had listed for aconitine poisoning. Despite the lack of certainty about the vehicle, Dr. Lamson was found guilty by the jury after only about 30 minutes of deliberation. In the Agatha Christie version [46], the poison was in a raisin in the slice of 'Dundee' cake given to Percy. Dr. Lamson was still addicted to morphine when sent to prison; he confessed four days before he was executed for the murder of Percy John.

NOTES

[1] Winston Churchill (1874-1965). In Consuelo Vanderbilt Balsan, *The glitter and the Gold*, 7, 1952.

[2] *Ode on Melancholy*. John Keats (1795-1821). The Oxford Book of English Verse: 1250-1900. Arthur Quiller-Couch, ed. 1919.

REFERENCES

[1] Poisonous Plants of the United States [Internet]. [cited 2017 May 9]. Available from: https://captainjamesdavis.net/2014/02/18/poisonous-plants-of-the-united-states/

[2] 7 of the World's Deadliest Plants | Britannica.com [Internet]. [cited 2017 May 9]. Available from: https://www.britannica.com/list/7-of-the-worlds-deadliest-plants

[3] The Green Killers: Poisonous Plants in History | Plan(e)t [Internet]. [cited 2017 May 13]. Available from: https://cambridgeplanet.wordpress.com/2014/11/22/the-green-killers-poisonous-plan-s-in-history/

[4] Where Does Nightshade Grow? | Home Guides | SF Gate [Internet]. [cited 2017 May 12]. Available from: http://homeguides.sfgate.com/nightshade-grow-82116.html

[5] Largo M, Bauer M. The big, bad book of botany. William Morrow 2014.

[6] Macbeth: About | CliffsNotes [Internet]. [cited 2017 May 12]. Available from: https://www.cliffsnotes.com/literature/m/macbeth/about-macbeth

[7] Executed in Error | Hawley Crippen | Secrets of the Dead | PBS [Internet]. [cited 2017 May 8]. Available from: http://www.pbs.org/wnet/secrets/hawley-crippen/199/

[8] Hyoscine | definition of hyoscine by Medical dictionary [Internet]. [cited 2017 May 12]. Available from:http://medical-dictionary.thefreedictionary.com/hyoscine

[9] Steele J. A Brief History of the Use of Poisonous Plants [Internet]. [cited 2017 May 5]. Available from: http://www.manyhatspublications.com/article-144.html

[10] Doctor - Biography.com [Internet]. [cited 2017 May 8]. Available from: http://www.biography.com/people/hawley-crippen-17172114

[11] Young F. The trial of Hawley Harvey Crippen, ed. with notes and an introduction by Filson Young [Internet]. [cited 2017 May 10]. Available from: https://archive.org/details/trialofhawleyhar00cripiala

[12] Emsley J. Molecules of Murder: Criminal Molecules and Classic Cases [Internet]. [cited 2017 May 9]. Available from: https://books.google.com/books?id=JAzKSP4NagQC&pg=PA47&lpg=PA47&dq=murder+by+belladonna&source=bl&ots=q-RzNzhCH2&sig=iELPfRhkEYAEJsn1wOIi6himXI0&hl=en&sa=X&ved=0ahUKEwiInPyuvuHTAhWKwVQKHb61DdYQ6AEIfzAR#v=onepage&q=murder

[13] Cavendish R. 2010. The Execution of Dr Crippen [Internet]. History Today. 2010 [cited 2017 May 8]. Available from: http://www.historytoday.com/richard-cavendish/execution-dr-crippen

[14] Boyne J. Crippen : a novel of murder. Black Swan 2011.

[15]   Hawley Harvey Crippen, Miscarriage of Justice. [Internet]. [cited 2017 May 8]. Available from: http://www.defrostingcoldcases.com/hawley-harvey-crippen/

[16]   Poe C, Clemens R. A Study of Vitali's Test for Atropine J Lab Clin Med [Internet] Elsevier , 1933 [cited 2017 May 11];18(7): 743-50. Available from: http://www.translationalres.com/article/ S0022-2143(33)90446-1/abstract

[17]   The Destruction Of Alkaloids By Emulsions Of The Body Tissues on JSTOR [Internet]. [cited 2017 May 11]. Available from: https://www.jstor.org/stable/25299085?seq=2#page_scan_tab_contents

[18]   1918. A MONOGRAPH ON TREMBLES OR MILKSICKNESS AND WHITE SNAKEROOT - TECHNICAL BULLETIN 15 - JULY, 1918: F. A. ; R. S. Curtis; B. F. Kaupp Wolf: Amazon.com: Books [Internet]. [cited 2017 May 17]. Available from: https://www.amazon.com/MONOGRAPH-TREMBLES-MILKSICKNESS-WHITE-SNAKEROOT/dp/B004G7HR54

[19]   Davis TZ, Stegelmeier BL, Lee ST, Collett MG, Green BT, Pfister JA, *et al.* 2015 [cited 2017 May 17];106: 29-36. White snakeroot poisoning in goats: Variations in toxicity with different plant chemotypes. YRVSC [Internet]. Available from: http://ac.els-cdn.com.proxy.mul.missouri.edu/ S0034528816300315/1-s2.0-S0034528816300315-main.pdf?_tid=5d136582-3b38-11e--8afb-00000aab0f6b&acdnat=1495050072_629834ec7b807ca4df4212e362f905fe

[20]   Jordan P. Indiana magazine of history [Internet] Indiana Magazine of History Indiana University, Dept of History] , 1944 [cited 2017 May 12]; Available from: https://scholarworks.iu.edu/journals/ index.php/imh/article/view/7476/8665

[21]   White Snakeroot | MDC Discover Nature [Internet]. [cited 2017 May 12]. Available from: https://nature.mdc.mo.gov/discover-nature/field-guide/white-snakeroot

[22]   1809. The Kentucky Encyclopedia - John E. Kleber - Google Books [Internet]. [cited 2017 May 12]. Available from: https://books.google.com/books?id=8eFSK4o--M0C&pg=PA637&lpg=PA637&dq= %22milk+sickness%22+1809&hl=en#v=onepage&q=%22milk

[23]   Doctor Anna, Anna Pierce Hobbs Bigsby,1808-1869 , 1998 [cited 2017 May 17]; Available from: http://www.alliancelibrarysystem.com/IllinoisWomen/files/lg/html/lg000003.html

[24]   Nancy Hanks Lincoln - Lincoln Boyhood National Memorial (U.S. National Park Service) [Internet]. [cited 2017 May 17]. Available from: https://www.nps.gov/libo/learn/historyculture/nancy-hank--lincoln.htm

[25]   Milk Sickness - Abraham Lincoln Birthplace National Historical Park (U.S. National Park Service) [Internet]. [cited 2017 May 17]. Available from: https://www.nps.gov/abli/planyourvisit/ milksickness.htm

[26]   The Two Mothers Who Molded Lincoln - History in the Headlines [Internet]. [cited 2017 May 17]. Available from: http://www.history.com/news/the-two-mothers-who-molded-lincoln

[27]   Davies E. Some Common Species Ranked as the Deadliest Plants in the World. [Internet]. [cited 2017 May 12]. Available from: http://www.bbc.com/earth/story/20150817-earths-most-poisonous-plants

[28]   Glukman E. Castor Bean and Jequirity Bean Poisoning: Background, Pathophysiology, Epidemiology [Internet]. [cited 2017 May 12]. Available from: http://emedicine.medscape.com/article/ 1009200-overview

[29]   Falach R, Sapoznikov A, Gal Y, Israeli O, Leitner M, Seliger N, *et al.* Quantitative profiling of the in vivo enzymatic activity of ricin reveals disparate depurination of different pulmonary cell types Toxicol Lett [Internet] , 2016 [cited 2017 May 13];258: 11-9. Available from: http://www.sciencedirect.com/ science/article/pii/S0378427416301436

[30]   Umbrella Assassin | Full Episode | Secrets of the Dead | PBS [Internet]. [cited 2017 May 15]. Available from: http://www.pbs.org/wnet/secrets/umbrella-assassin-watch-full-episode/1549/

[31]   Hignett K. Vladimir Kostov « The View East [Internet]. [cited 2017 May 15]. Available from: https://thevieweast.wordpress.com/tag/vladimir-kostov/

[32]   Abrus precatorius – UF/IFAS Center for Aquatic and Invasive Plants [Internet]. [cited 2017 May 13]. Available from: https://plants.ifas.ufl.edu/plant-directory/abrus-precatorius/

[33]   Kumar MS, Karande AA. A monoclonal antibody to an abrin chimera recognizing a unique epitope on abrin A chain confers protection from abrin-induced lethality. Hum Vaccin Immunother [Internet] Taylor & Francis , 2016 Jan 2; [cited 2017 May 15];12(1): 124-31. Available from: http://www.tandfonline.com/doi/full/10.1080/21645515.2015.1067741 [http://dx.doi.org/10.1080/21645515.2015.1067741]

[34]   Sabo T, Gal Y, Elhanany E, Sapoznikov A, Falach R, Mazor O, *et al.* Antibody treatment against pulmonary exposure to abrin confers significantly higher levels of protection than treatment against ricin intoxication. Toxicol Lett [Internet] , 2015 [cited 2017 May 15];237(2): 72-8. Available from: http://www.sciencedirect.com/science/article/pii/S0378427415002131 [http://dx.doi.org/10.1016/j.toxlet.2015.06.003]

[35]   Mishra R, Kumar MS, Karande AA. Inhibition of protein synthesis leading to unfolded protein response is the major event in abrin-mediated apoptosis. Mol Cell Biochem [Internet] Springer US , 2015 May 10; [cited 2017 May 15];403(1): 255-65. Available from: http://link.springer.com/10.1007/s11010-015-2355-9 [http://dx.doi.org/10.1007/s11010-015-2355-9]

[36]   Bhasker ASB, Sant B, Yadav P, Agrawal M, Lakshmana Rao PV. Plant toxin abrin induced oxidative stress mediated neurodegenerative changes in mice. Neurotoxicology [Internet] , 2014 [cited 2017 May 15];44: 194-203. Available from: http://www.sciencedirect.com/science/article/pii/S0161813X14001065 [http://dx.doi.org/10.1016/j.neuro.2014.06.015]

[37]   Chaturvedi K, Jadhav SE, Bhutia YD, Kumar O, Kaul RK, Shrivastava N. Purification and dose-dependent toxicity study of abrin in swiss albino male mice. Cell Mol Biol 2015; 61(5): 36-44. [PMID: 26475386]

[38]   M is for Monkshood | BizarreVictoria [Internet]. [cited 2017 May 16]. Available from: https://bizarrevictoria.wordpress.com/2017/02/06/m-is-for-monkshood/

[39]   Rare Flower Discovered in Ozarks - Ozark National Scenic Riverways (U.S. National Park Service) [Internet]. [cited 2017 May 13]. Available from: https://www.nps.gov/ozar/learn/nature/monkshood.htm

[40]   Gardener "died after brushing past poisonous plant" in millionaire's garden - Telegraph [Internet]. [cited 2017 May 13]. Available from: http://www.telegraph.co.uk/news/uknews/law-and-order/11213530/Gardener-died-after-brushing-past-poisonous-plant-in-millionaires-garden.html

[41]   Monkshood - AACC.org [Internet]. [cited 2017 May 15]. Available from: https://www.aacc.org/community/divisions/tdm-and-toxicology/toxin-library/monkshood

[42]   Herbs-Treat and Taste: ACONITE, POISONOUS PLANT: HISTORY OF USES OF ACONITE NAPELLUS [Internet]. [cited 2017 May 17]. Available from: http://herbs-treatandtaste.blogspot.com/2012/02/aconite-poisonous-plant-history-of-uses.html

[43]   Cerberus | Pitlane Magazine [Internet]. [cited 2017 May 17]. Available from: http://www.pitlanemagazine.com/cultures/cerberus.html

[44]   Cerberus in Greek Mythology - Greek Legends and Myths [Internet]. [cited 2017 May 17]. Available from: http://www.greeklegendsandmyths.com/cerberus.html

[45]   2015. ADAM HL. TRIAL OF GEORGE HENRY LAMSON (CLASSIC REPRINT). [Internet]. FORGOTTEN BOOKS; 2015 [cited 2017 May 16]. Available from: https://www.forgottenbooks.com/en/books/ TrialofGeorgeHenryLamson_10298222

[46]   Harkup K. Harkup K. A is for arsenic : the poisons of Agatha Christie [Internet]. [cited 2017 May 17]. 320 p. Available from: http://www.bloomsbury.com/us/is-for-arsenic-9781472911308/

CHAPTER 5

Cyanide

Abstract: The art and history of cyanide as a poison begins innocently with the discovery of Prussian blue as a welcome addition to the palette of artists and dyers of cloth where a vivid blue was difficult to obtain and expensive. It did not take long for the dark and sinister uses of this poison to emerge and continue. A form of cyanide may have been used by the Emperor Nero to murder members of his family. It probably was added to the bayonets of soldiers of Napoleon III in the Franco-Prussian War. In World War II, unspeakable horrors were committed by the Nazis who used Zyklon B (cyanide) to commit mass murders in death camps. Near the end of that war, cyanide was used in mass suicides by the German people, and infamous Nazis, including Hitler, Eva Braun, the Goebbels, and others committed suicide with it. A small quantity of cyanide, taken internally kills almost instantly. Enough cyanide to kill could be placed in a glass capsule that could be conveniently hidden on a person's body or even secreted in the mouth to be crushed when death by suicide was the chosen option. The mass suicide by more than 900 members of the Jim Jones cult was caused by a drink laced with cyanide.

"Wickedness is always easier than virtue, for it takes the shortcut to everything."[1]

Keywords: Arsenic, Berlin Blue, Carcinogen, Cyanide, Cytochrome a_3, Demmin, Drinking Water, Gainsborough, Hitler, Hypoxia, Jim Jones, King of Poisons, Mees' Lines, Nero, Picasso, Prussian Blue, Rhodanese, Rommel, Starbuck Strawberry Frappuccino, Third Reich, Van Gogh, WW II, Zyklon B.

INTRODUCTION

The art of use and the history of cyanide began in the eighteenth century with the accidentally-on-purpose creation of Prussian blue. The story begins like this. Heinrich Diesbach was an artist but also a scientist. It is said that he devoted hours to experimentation in the Berlin laboratory of a friend trying to create a new red paint for his artist's palate [1]. Eventually in 1704 he heated blood, potash and green vitriol and let it simmer. Potash is a water-soluble form of potassium and green vitriol is reduced iron combined with sulfate. Presumably, he expected to get something useful in the form of a red pigment that he could use in creating his oil paintings. Instead, he obtained a deep violet-blue chemical substance that he called Berlin Blue. English chemists later renamed it Prussian blue, the name that has endured. Nearly 80 years later a Swedish chemist mixed Diesbach's pigment with acid and a colorless gas with the smell of bitter almonds was emitted [1]. The gas when condensed becomes the clear liquid known as Prussic acid. Prussic acid is one of the best sources to generate the poisonous gas cyanide.

Diesbach's discovery was a combination of serendipity and perseverance [2]. He was trying to create something called 'cochineal red lake'. The term 'lake' is still used to connote a liquid form of a colored substance that is useful in oil painting. Cochineal referred to the natural dye carmine obtained by crushing the bodies of 'scale' insects. These insects live on prickly pear cacti (which was common to the area where I grew up). The insects eat red parts of the cactus and concentrate the color in their bodies. Colored pigments derived from these insects are commonly used in the food industry, and a 'dust-up' occurred with Starbucks strawberry frappuccinos reported by CBS News in 2012 [3]. It seems their desire to use natural rather than artificial coloring had resulted in Starbucks switching to cochineal 'bugs' which caused concerns among vegans.

Back to Diesback, with details provided by Marion Boddy-Evans [2]. He is said to have tried to save money by buying some cheap ingredients that included contaminated potash. Mixing his ingredients, he expected to get a deep red color. Instead he got only a pale-colored substance which he tried to concentrate. However, instead of deep red he obtained the deep blue we described earlier. This account [2] says that ultramarine was made from lapis lazuli ("blue stone"), the principal blue pigment at that time, but which was only available from Badakshan (now Afghanistan) and was more expensive than gold. This account says that Diesbach made his discovery at some time between 1704 and 1705. By 1710 it was described as 'equal to or excelling ultramarine'. In fact, it was about a tenth the price and by 1750 Prussian blue was in wide use across Europe. Famous artists including Gainsborough (Blue Boy), Constable, Monet, Van Gogh, and Picasso (famously noted for his blue period) painted with it.

Paintings before the eighteenth century rarely have blue and when found it mostly occurs in paintings of religious subjects and even then it is most used in the clothing of people considered to be very saintly [4]. As stated, the pigment ultramarine was most used to create blue in paintings and ultramarine was made from the mineral lapis lazuli. It is said that the name ultramarine, meaning 'beyond the sea' was appropriate became it came from Afghanistan. This blue color was expensive and rare [4]. The discovery of Prussian blue changed this. It was very stable, even when exposed to strong light, and it was much cheaper than any other blue paint. For art lovers, the introduction of Prussian blue is both a way of dating artwork and experiencing how art and science (even primitive science, hardly out of the dark ages) come together.

The chemical structure of Prussian blue is complex and contains iron in both the oxidized and reduced form with six attached cyanide groups (Fig. **1**). The chemistry resulting from the complex arrangement of cyanide in the molecule results in a structure with reduced iron and oxidized iron existing in slightly different environments. Light, striking the pigment, is selectively absorbed according to its wavelength. Orange light, transfers an electron from the reduced iron to oxidized iron. Chemically, this is known as charge transfer: an electron (negatively charged) moves from one molecular orbital to another in a different atom (iron in this case). The energy necessary to make this transfer comes from the light absorbed. However, it is the unabsorbed light which is the opposite color (blue) that we see by reflection.

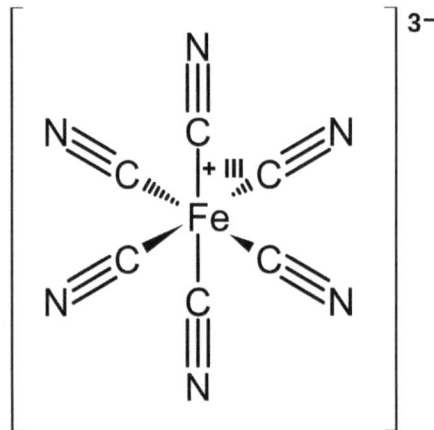

Fig. (1). Part of the chemical structure of Prussian blue showing iron +3 (+III) surrounded by 6 cyanide functional groups. See text for more details.

Prussian blue is potentially a bit dangerous since under unusual conditions cyanide gas can be released. It also has an important medical use – the treatment of patients contaminated internally with certain metals such as radioactive cesium or the highly poisonous thallium [4]. Prussian blue is given orally and absorbs the metals which are then excreted. With this treatment the time spent in the body by radioactive cesium is reduced from 110 days to approximately a month and the time spent in the body by ingested thallium is reduced from about 8 days to about 3 days [5]. The toxicity is thus decreased.

CYANIDE, THE POISON

Brief History

The chapter 'Cyanide' in the book Physician's Guide to Terrorist Attack [6], is a useful source (with many original source references) of information summarized here. The authors state that cyanide and plant materials containing cyanide have been used since antiquity. Cyanide was first isolated chemically by Scheele from Prussian blue in 1782 [7]. There is uncertainty about whether it had a role in Scheele's death. Scheele experimented with it not knowing of its poisonous nature.

Cyanide has been associated with murders and assassinations. It was an ineffective chemical weapon in WWI. It was used by the German Third Reich in mass exterminations of prisoners in death camps. The latter use was as Zyklon B, the trade name of a cyanide-based pesticide developed in Germany in the early part of the 1920s. More about this tragic story will be included later in this chapter.

Technically, cyanide refers to any chemical containing a carbon-nitrogen (CN) bond. Cyanide has carbon triply-bonded to nitrogen and has a charge of minus one. Some similar compounds are deadly poisons; others (some of which are called nitriles) are not. The CN⁻ ion, which is the metabolic poison in cyanide toxicity, is not readily formed by nitriles. A nitrile is found, for example in the common over-the counter antacid Tagamet [8]. This reference also has a long list of sources of low levels of exposure to cyanide that includes eating cassava, lima beans, yucca, almonds and apple seeds. In apples the cyanide is in the seeds which are sometimes consumed. In cherries, apricots and peaches the cyanide is in the pits which of course are not consumed. People who smoke cigarettes are exposed to cyanide. Nonsmokers average about 0.06 micrograms of cyanide per ml of blood but smokers have about 0.17 according to a reported cited by Baskin and Brewer [9]. There is a possible danger from inhaling smoke from burning of plastics containing nitrogen. Indeed, the combustion of any substance that contains carbon and nitrogen has the potential to form cyanide. Industrial production of cyanide in the United States amounts to more than 300,000 tons of hydrogen cyanide annually [10].

The military uses of cyanide as a weapon are limited. Although inhalation of cyanide gas is rapidly lethal (large doses can kill in 6 to 8 minute), it rapidly disperses in the environment and cyanide requires a relatively large concentration in air to be lethal. Therefore, large munitions (such as artillery shells, for example) are required. Even then, upon explosion cyanide is rapidly dispersed to non-lethal air concentrations. Cyanide is highly volatile and skin decontamination after exposure is not usually required [10]. Antidotes are effective if administered quickly [9]. It is said that about 4,000 tons of cyanide was used by the French in WWI without notable military success "possibly because the small one- to two-pound munitions used could not deliver the large amounts needed to cause biological effects" – and it causes few effects below its lethal concentration. Conversely, mustard gas causes eye damage at 1% of the lethal concentration [10]. It is said that the U.S. had a small number of cyanide munitions during WW II; Japan allegedly used cyanide against China before and during WWII; and Iraq may have used cyanide against the Kurds in the 1980s [10]. It is fortunate that this evil use of cyanide was thwarted by the natural properties of cyanide.

Considerable information about military uses of cyanide, with source references, is found in the book by Baskin and Brewer [9]. They relate that Nero, in Roman times, used cherry laurel water, which contains cyanide as its chief toxic component, to poison some of his family and other people out of political favor with him. Napoleon III allegedly proposed cyanide for use on his armies' bayonets during the Franco-Prussian War. In the early parts of WWI, the French proposed using cyanide and hydrocyanic acid. The latter was a poor choice for this evil purpose because hydrocyanic acid persists in air for only a very few minutes and the Germans quickly equipped their troops with a mask that was protective. By the fall of the year 1916, the French tried cyanogen chloride, a heavier and less volatile substance, for their nefarious purpose. This gas had cumulative effects on victims that included eye and lung irritation which, unfortunately for the victims, required only comparatively low concentrations. At high concentrations, this gas rapidly paralyzed the respiratory system. Near this time, the Austrians began to use a poisonous gas (cyanogen bromide) that was both less volatile and less deadly than hydrocyanic acid. However it was found to be useful because it had a strong irritating effect on the eyes and the entire respiratory system including the lungs. However, it proved to be corrosive to metal storage containers which, of course, could not be tolerated. It also readily reacted to become detoxified and Austria abandoned it [9].

In WWII, Nazi Germany killed millions in death camps with a formulation of hydrocyanic acid (Zyklon B) that initially had been developed as a fumigant and rodenticide. Zyklon B (there had been a forerunner Zyklon A) was used prior to and during WWII for disinfection and extermination of pests aboard ships [9].

After the Nazi invasion of the Soviet Union, *Einsatzgruppen* (mobile killing squads) that followed the Nazi army rounded up and murdered large numbers of Jewish people by mass shootings. This included the horrors at Babi Yar. Based on historical accounts, this atrocity was deemed too costly, too slow and most importantly to them, it had a devastating mental effect on the soldiers who did the shootings [11]. This selective humanitarian concern drove those in charge to seek more 'efficient' means of mass killing of those they deemed unworthy of life.

In the Auschwitz concentration camp, Zyklon B (Fig. **2**) was used for sanitation and pest control until the summer of 1941 [12]. Beginning late in August, 1941 Zyklon B was used, first experimentally and then routinely, as an agent of mass annihilation. Jennifer Rosenberg, in a web page [11] records that:

> "Rudolf Höss, the commandant of Auschwitz and Adolf Eichmann were searching for a more expedient way to exterminate large numbers at the same time when they decided to try Zyklon B. On September 3, 1941, 600 Soviet prisoners of war and 250 Polish prisoners who were no longer able to work were forced into the basement of Block 11 at Auschwitz I, known as the 'death block', and Zyklon B was released inside. All died within minutes."

Fig. (2). Canisters of Zyklon B.

Zyklon B consisted of diatomite (a form of diatomaceous earth) in granules the size of small peas that were saturated with prussic acid, and sealed in metal canisters. It was produced by two German firms (*Tesch/Stabenow* and *Degesch*) who obtained the patent from the giant company *IG Farbenidustrie AG* [13]. This report [13] states that at some point *Tesch* supplied two tons a month and *Degesch* three quarters of a ton of Zyklon B. These companies had experience with the use of cyanide as a fumigant and as a pesticide to exterminate rodents. After WWII was over, the directors of these firms insisted they had sold Zyklon B for fumigation purposes and knew nothing about it being used to kill humans. Prosecutors at the Nuremburg trials found letters from *Tesch* offering to supply 'gas crystals' but also advising about how to use certain ventilating and heating equipment. There also was other damning testimony. Records showed that they had sold enough to kill two million people. This was most difficult to dismiss. Two men associated with *Tesch* were sentenced and hanged in 1946. The directors of *Degesch* were sentenced to five years in prison [13]. This web page

[13] provides a gruesome account of the use of Zyklon B (Fig. **3**). Hans Stark, registrar of new arrivals at Auschwitz at the time, is reported to have recounted:

Fig. (3). An example of one of many reinforced concrete chambers used to 'exterminate' prisoners with Zyklon B.

"At another, later gassing– also in autumn 1941– Grabner [Maximillian Grabner, said to be "Head of the Political Department at the Auschwitz"] ordered me to pour Zyklon B into the opening because only one medical orderly had shown up. During a gassing Zyklon B had to be poured through both openings of the gas-chamber room at the same time. This gassing was also a transport of 200-250 Jews, once again men, women and children. As the Zyklon B - as already mentioned - was in granular form, it trickled down over the people as it was being poured in. They started to cry out terribly for they knew what was happening to them. I did not look through the opening because it had to be closed as soon as the Zyklon B had been poured in. After a few minutes there was silence. After some time had passed, it may have been ten to fifteen minutes, the gas chamber was opened. The dead lay higgledy-piggledy all over the place. It was a dreadful site."

This is quoted in its entirety as there is no possible way to do otherwise for this horrific account. Even though cyanide kills very quickly and incapacitates even more quickly, in mass gassings the effect is most horrific and is heightened by the odor, the observed effect on others, and the futility of no escape.

Cyanide, by most accountings, is the least toxic of the other 'war gases' that have been made in the laboratory. This is a gruesome testament to the assiduity of man. The lethal concentration of hydrocyanic acid in air that kills 50% of humans by inhalation is estimated to be 2,500 to 5,000 milligram minutes per cubic meter of air, and approximately 11,000 for cyanogen chloride (the forms of cyanide used as chemical airfare agents). The lethal dose for hydrogen cyanide by intravenous injection is estimated to be 1.1 mg per kg of body weight and 100 mg/kg of body weight *via* skin exposure. The oral LD 50 dose is reported to be 100 mg/kg of body weight for sodium cyanide and 200 mg/kg of body weight for Potassium cyanide [10]. Cyanide is said to be unique among the chemical war agents because it is detoxified by the human body at a significant rate: 17 femtagrams per kg of body weight. This means that to be lethal, the exposure time becomes important. The effect of exposure time on lethality is not directly proportional to the concentration in the air [10]. The small amount of continual metabolism of cyanide that occurs in the body occurs primarily by the liver enzyme rhodanase (also called rhodanese). Rhodanase catalyzes the irreversible reaction of cyanide to the thiocyanate ion which is comparatively non-toxic and is excreted. This reaction is sufficient to detoxify the small amounts of cyanide formed in the body from various 'natural' sources but it is overwhelmed when poisoning occurs.

CLINICAL EFFECTS OF CYANIDE

Cyanide Poisoning

The signs and symptoms of acute cyanide poisoning and treatment options have been reviewed in two recommended publications which together contain more than 100 cited references [14, 15]. The following accounts are drawn primarily from these reviews which can be consulted for original sources.

Cyanide poisoning in societies today results from various household, industrial, and natural exposures. Smoke inhalation from burning buildings is the most common cause in the Western World [15]. Natural material such as silk and wool, and man-made polymers can produce cyanide gas during thermal degradation. Although carbon monoxide may be the primary culprit in such cases, cyanide has a significant role [15].

Medical, intravenous administration of sodium nitroprusside as an anti-hypertensive also can generate cyanide [15]. Cyanide is formed in small amounts in the body non-enzymatically. However, as previously stated, conversion of small amounts in the liver by the enzyme rhodanase safely detoxifies this naturally, and the product is excreted. However, when therapy for high blood pressure by this means is extended, especially if there is kidney impairment, cyanide poisoning can unfortunately result.

Mechanisms of Cyanide Toxicity

Cyanide results in tissue anoxia by binding to cytochrome oxidase a_3 in mitochondria. Mitochondria are subcellular organelles that contain the respiratory machinery where oxidative phosphorylation occurs. This oxidative metabolic process connects electrons, stripped from oxygen, to the enzymatic oxidation of chemicals derived from foods. This process produces the energy required by every cell. The energy is 'stored in the chemical called adenosine triphosphate (ATP). Cytochrome a_3 is the final step in this energy-storing process. The brain uses approximately 25% of the glucose consumed at rest by the human body, and poisoning of the production of ATP impairs the brain so rapidly that signs and symptoms can appear within one minute by inhalation or within a few minutes by ingestion of lethal amounts of cyanide [15].

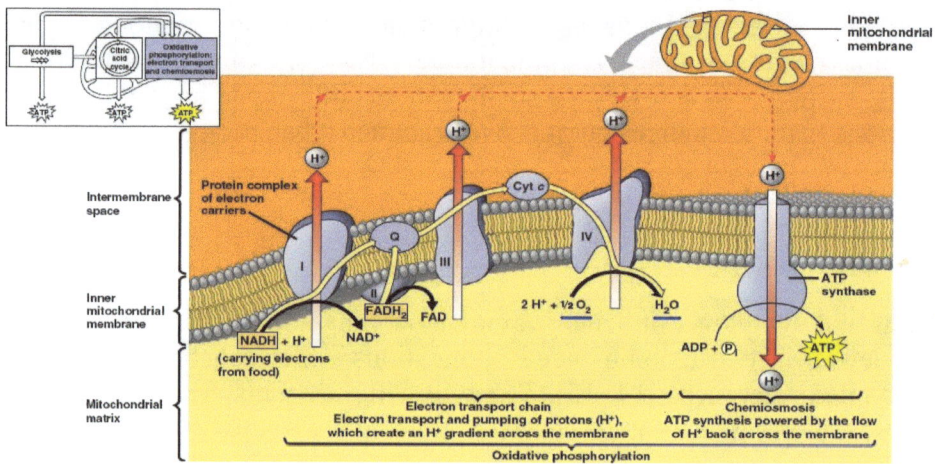

Fig. (4). Intracellular site of cyanide poisoning (see text for details).

Cyanide specifically binds to ferric iron which is part of the cytochrome oxidase a_3 enzyme within mitochondria. It blocks an essential step (III to IV) in the production of ATP (Fig. **4**). The binding by cyanide is reversible but it effectively stops cellular respiration by blocking the reduction of oxygen to water which is required as the final step in oxidative phosphorylation. As previously stated, cyanide is detoxified in the body. The enzyme rhodanase converts cyanide to nontoxic thiocyanate in the liver and it is excreted by the kidneys. The low-levels of cyanide naturally occurring in the body are detoxified in this manner; however, the mechanism is overwhelmed with large doses of cyanide.

Cyanide has considerable affinity for the ferric (oxidized) iron of methemoglobin. This results in a therapy that induces methemoglobin formation with resultant preferential binding of cyanide way, thus sparing the cytochrome a_3 site [10]. Methemoglobin is formed in the blood from hemoglobin.

Cyanide is a composed of carbon triply-bonded to nitrogen (a cyano group) and other positively charged elements are associated (usually potassium, sodium, or hydrogen (Fig. **5**). Cyanide exists as a salt, a liquid, or a gas. In the body it reacts quickly with metals including oxidized (3^+ iron) and it blocks enzymes that lead to a cascade of toxic events that can be lethal.

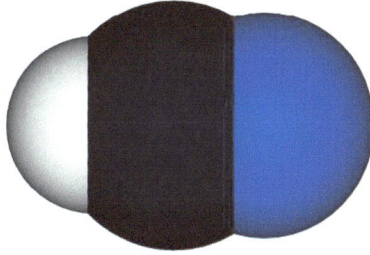

Fig. (5). Space-filling model of cyanide which has a carbon atom (black) triply-bonded to nitrogen (blue) with another atom such as hydrogen (silver-white) or group of atoms also bonded to carbon, or it is negatively charged.

Technically, the term cyanide refers to both the anion CN^- and its acidic form hydrocyanic acid (HCN) [10].

Signs and Symptoms of Cyanide Poisoning

With life-threatening poisoning there is rapid hypoxia with almost immediate anxiety, headache, inability to focus the eyes and dilation of the pupils. As hypoxia progresses rapidly to anoxia there is loss of consciousness and seizures and coma can result in death [15].

The clinical effects of cyanide stem primarily from the consequences of intracellular hypoxia. "The onset of signs and symptoms is usually less that 1 minute after inhalation and within a few minutes after ingestion" [15] of significantly toxic amounts of cyanide. Early signs of cyanide poisoning are transient rapid and deep respirations from stimulation of peripheral and central chemoreceptors in the brain stem that is the body's response to tissue hypoxia (low oxygen concentration). Hypoxia profoundly affects the heart and cardiac output may be initially increased, and blood pressure may increase as a result of catecholamine release. Vasodilatation and hypotension ensues with a shunting of blood to the brain and heart. Depression of the sino-atrial node, an increase in arrhythmias and decrease in the heart contraction force leads to cardiac arrest and death [15].

Serum concentrations of cyanide greater than 0.5 mg/liter are typically found with acute cyanide poisoning as cited by a reference found in the review by Jillian

Hamel [15]; however, cyanide concentrations, oddly, do not correlate well with severity of poisoning. This review also states that "Serum lactate levels greater than 8 millimole per liter are associated with acute poisoning and may aid in determining the need for repeated antidotal therapy". However, it should be noted that increased lactate is not specific to cyanide poisoning.

Treatment for Cyanide Poisoning

Even life-threatening poisoning by cyanide is potentially treatable when quickly diagnosed, but any delay likely will lead to hypoxic brain injury and cardiovascular impairment rapidly leading to death [15]. Antidotes ideally should be administered as soon as possible; however, there is no reliable test to confirm cyanide poisoning.

The 33rd annual report of the National Poison Data System of the American Association of Poison Control Centers for 2015 documented 205 cases of cyanide exposure in the United States with 86 patients treated in health care facilities. The number of outcomes by categories included: 28 were minor, 6 were moderate, 6 were major, and 3 resulted in deaths [16]. A total of 2,168,371 'encounters' for humans were logged and 7,657 were said to be confirmed non-exposures.

CYANIDE POISONINGS, THE UNUSUAL AND THE BIZZARE

The rapid physiological responses (previously described) provide an understanding of why cyanide pills have been developed as a means of suicide and have been used notoriously to prevent capture and its consequences by historical figures including Adolph Hitler and Eva Braun, Goering, Himmler and his wife (and they to their children).

The Circumstances of Rommel's Death by Cyanide

Erwin Rommel, a military commander in the Nazi regime in WWII was accused late in 1944 of involvement in a failed assassination plot against Adolf Hitler. Hitler was enraged and there was a wide-sweeping effort for revenge against all who were involved. History has not resolved the extent, if any, of Rommel's participation in the plot but it is known that, loyal to Hitler initially, Rommel had favored a political effort to remove Hitler near the end of the war. On 14 October 1944, two Nazi henchmen came to Rommel's home and told him that he (Rommel) could either commit suicide or face a trial for treason. The treason trial would have death as its certain outcome for himself plus humiliation and likely risk of death to his family. Rommel chose death by a cyanide pill [17, 18].

Rommel is remembered differently by historians than most who served Hitler in the Third Reich. He was both an honorable soldier, and a fierce adversary to the Allied Forces. By most accounts, he had no involvement in the death camp atrocities, and probably no involvement in the plot to kill Hitler. He was a loyal supporter of Germany and had served throughout WWI, taking part personally in missions with great personal danger. He was wounded at least twice and his unit captured about 9,000 Italian soldiers. In WWII he was known as the "Desert Fox" for his cunning in tank battles in North African. It is a matter of historical record that British Prime Minister Winston Churchill praised Rommel for his daring and skill in speeches in the House of Commons. General George Patton respected his skill in mobile, tank warfare. In 1944, Rommel was placed in charge of overseeing the massive efforts to strengthen the 'Atlantic Wall' – the name given to the extensive German coastal defenses against invasion. Soon after the successful D-Day invasion, an allied aircraft strafed Rommel's car causing it to crash. Rommel was severely injured with multiple skull fractures and other injuries to his face. This serious injury to Rommel was used officially to 'explain' his death and because of his popularity with the German people, he was given a state funeral [17]. Unlike most Germans who had significant roles in Hitler's failed Nazi Regime, Rommel's name post-WWII has not been vilified.

Rommel's death was certain upon crushing the cyanide pill. His son, Manfried Rommel was 15-years-old at the time. He provided a type-written account of his father's death on April 27, 1945 in what apparently was a dictated account because it was typed in English. This poignant account is included in a page available on-line [19]. It tells us much about the relationship between Rommel and his son, about the terrible time in Germany at the end of WWII, and about a death by cyanide that was seen as a preferable alternative to other choices available. The typed account by Manfried can best be presented as direct quotes [19].

> "My father was well, and we had breakfast together. We went for a walk… two Army Generals would be coming that day… the affair struck him as somewhat suspicious, and that he was not sure whether the reason given, to discuss his future appointment, was not being made to serve as a camouflaged plot for his removal. At 12'o'clock my father received the two Generals. My father asked me to leave the room… I met my Father just coming out of my Mother's room… He told me that he had taken his leave of my Mother, and that Adolf HITLER had given him the choice between taking poison or being brought before the People's Court. Adolf HITLER had also let him know that in the event of his committing suicide, nothing was to happen to his family, the family, on the contrary, would be provided for." [Capitals in the original].

A two-page document including this account surfaced after being sold at auction.

It discloses that Rommel left the house in uniform; was saluted by the two generals and left in the car with them. Approximately 15 minutes later a telephone call to Rommel's wife came from a hospital saying: "… my father had been brought there and apparently succumbed to an attack of cerebral apoplexy". Manfried Rommel had planned on joining the notorious Waffen SS but because of his father's opposition he joined the Luftwaffe. Manfred Rommel survived WWII, became a lawyer and mayor of Stuttgart [19]. He passed away on November 7, 2013. If there is an honorable suicide, Rommel's self-poisoning by cyanide reasonably can be said to have been one, in my opinion. Poisons, like modern technology, are neither good nor evil, but their use determines any moral content.

Suicides of Hitler and Eva Braun

I have taken information from four accounts to focus on the use of cyanide as the means of suicide for Hitler and Eva Braun [20 - 23]. Eva Braun, actually was Eva Hitler at this time, she and Hitler were married the day before their deaths on April 30, 1945. One of the best sources of evidence regarding the suicide deaths is the account by Rochus Misch who was a Hitler bodyguard at the time. Misch is said to have been the first person to view Hitler's body and has written a book. Both Eva and Hitler took cyanide by crushing a pill in their mouths, and Hitler also shot himself in the head. After the shot was heard, apparently, this occurred: "The study door was opened and Misch looked inside… My glance fell first on Eva. She was seated with her legs drawn up, her head inclined towards Hitler… Near her… the dead Hitler. His eyes were open and staring, his head fallen forward slightly" [22]. Misch died in Berlin in 2013 at the age of 96.

Suicides of Herman Göring and other Nazi Leaders.

It is said that "Even though it was invented by the British and the Americans, the [cyanide] pill was really famous among their enemies, the Nazis" [20]. Heinrich Himmler, Head of the *Schultzstaffel* (SS) and a leading member of the Nazi Party, was caught by the British several days after Germany's official surrender. He was apprehended during his cowardly attempts to hide and was taken to a British interrogation Camp on May 23, 1945. He was brought before a medical examiner and refusing to open his mouth, he crushed a cyanide pill he had hidden in his tooth. It is said that he died within 15 minutes [20].

Herman Göring (Fig. **6**), infamous in many roles in Nazi Germany and before in WWI, tried to evade capture after the war but was arrested by American forces. At the time he was under threat of death for treason by Nazi Germany for his attempts to take control from Hitler near the end of the war. Göring was declared guilty of war crimes and sentenced to death by hanging by the International Military Tribunal at Nuremberg. On October 15, 1946, one day before he was to be hanged, Göring committed suicide with a cyanide pill. Although an American guard was suspected of supplying the pill to Göring, whether this is true is not known [20].

Fig. (6). Herman Wilhelm Göring was a German fighter pilot in WWI; an infamous head of the Luftwaffe in Hitler's Nazi regime of WWII and President of the Reichstag.

Mass Suicides in Germany at the End of WWII

Over a mere 3-day period between April 30th and May 2nd of 1945, when the outcome of WWII was certain, over 1,000 Germans in the German city of Demmin decided to kill themselves with cyanide rather than face what would come with defeat [24]. There is no way for us today to understand the tragedy. Germans had lived through WWI, the rise of Hitler and wild expectations prior to and during the early parts of WWII. Then came the defeat of Hitler's army in Africa. By the time of the Normandy invasion Germany's territory had shrunk and the Russian mass armies swarmed from the north and east. It is said that eight out of 10 Nazis who died were killed by the Red army and that the Russians lost 20 million people during the war, nearly half of which were civilians [24]. These were strong reasons why the Germans feared retribution at the hands of the Russians. History relates that many Soviet soldiers believed there were no innocent Germans: "If you have not killed at least one German a day, you have wasted that day", and "If you leave a German alive, the German will hang a Russian and rape a Russian woman. If you kill one German, kill another – there is nothing more amusing for us than a heap of German corpses." This was written by the Russian intellectual Ilya Ehrenburg [25].

Demmin, where the suicides occurred, was a village of about 16,000 people in the Pomeranian area, a region where anti-Semitism and pro-Nazi sentiments had flourished in the 1920s and 1930s, even before the rise of Hitler. The city's residents participated violently in *Kristallnacht* (the night of broken glass) when

Jewish establishments were widely vandalized and Jewish people were terrorized.

Demmin was vulnerable; the city was surrounded by rivers and the Nazis destroyed the bridges to stall the Soviet advance; however, this trapped the German citizens in Demmin. Other German cities also witnessed mass suicides including perhaps 7,000 in the large metropolis of Berlin. However, the 1,000 in the small city of Demmin in 3 days is astounding [25]. Anticipating the Soviet army, residents hung white flags from windows as signs of surrender. Some resistance also occurred and it is recorded that one teacher killed his wife and daughter before firing an anti-tank grenade and then was killed by a means not reported. One Soviet soldier was killed.

Atrocities began in Demmin that evening about the time radio broadcasts informed the citizens that Adolf Hitler was dead by his own hands. Soviet soldiers had started breaking into homes to loot whatever they desired that was of value including jewelry and alcohol which they consumed. Soviet soldiers began lighting fires and in just three days 80% of the city's buildings were damaged or destroyed. The most terrifying atrocity was rape [25].

It is recorded that suicides took many forms: people hanged themselves, slit their wrists, shot themselves, and others ingested poison. Some jumped into the *Peene* and *Tollanse* rivers. Nazi propaganda toward the end of the war, including by the newspaper *Völkischer Beobachter*, encouraged suicide [25]. After the last concert at the Berlin Philharmonic on April 12, 1945, members of the Hitler Youth carried basket of cyanide capsules which they distributed to the audience [25]. "Like in a cult, the mass suicides in Nazi Germany were in part a response to the shock of seeing a massive, inextricable lie come crashing down." [25]. For me, this is an inadequate explanation; no explanation provides understanding. However, deaths by cyanide poisoning would have been less traumatic than some of the other means described, particularly when families were killed by their own members.

Jim Jones and a Mass Killing by Cyanide

The horrible mass murder of 909 people with cyanide dispensed in a drink to followers of Jim Jones was clearly evil [26]. Jones was a cult leader who organized a settlement in Guyana and lured people there with the hope of a utopian life. There is evidence that cult leader Jim Jones began buying cyanide in 1976, two years before he ordered more than 900 of his followers to drink a liquid describe as punch (and sometimes as Kool-Aid) on November 18, 1978 [27, 28]. A quarter to a half pound of cyanide per month was sent to the camp in Guyana. Stories were told by a few who did survive the mass killing. The mass suicides occurred shortly after gunmen under the direction of Cult leader Jim Jones killed U.S. House of Representative member Leo Ryan and four others who had visited the camp on November 18, 1978.

Jones told his followers, members of his Peoples Temple Church, that the Guyanese Army would invade their settlement and kill them. He demanded that the parents kill their children and then take their own lives. Of the 909 people who died, 303 were children including toddlers and teenagers. Many were killed by those most loyal to Jones by squirting cyanide down their throats using syringes. CNN reported that Jones obtained a jeweler's license to obtain the cyanide which can be used (inadvisably) in procedures with gold. The settlement actually had a doctor and he reportedly wrote a memo to Jones: "Cyanide is one of the most rapidly acting poisons… I would like to give about two grams to a large pig to see how effective our batch is" [28]. A group of survivors, who escaped early on the day of the mass killings, included a woman (Leslie Wilson) who walked almost 30 miles through 'the jungle' to a town while carrying her 3-year-old son. Jones reportedly was found dead with a gunshot wound to the head. It seems unbelievable that a plan could be made and carried out to buy cyanide, mix it in a drink and coerce more than 900 people to drink it and die. There was evidence that the punch also contained the sedative valium [26].

Slobodan Praljak

On November 29, 2017, Slobodan Praljak killed himself by drinking cyanide poison from a small container in a courtroom just after his 20-year jail term was upheld by a judge for crimes in the city of Mostar during the Bosnian war of 1992-1995. Praljak was one of six former Bosnian Croat political and military leaders charged by the International Criminal Tribunal. He declared that he was not a criminal. It is reported that Praljak "… stood and raised his hand to this mouth, tipped his head back and appeared to swallow a glass of liquid" [29]. He was taken to a hospital where he died.

Thus, it appears that even today, cyanide poisoning can be a means of public suicide which points to the rapidity and finality of it as a poison and in some cases to the fact that cyanide can be surreptitiously self-administered.

NOTES

[1] Samuel Johnson (1709-1784). 17 September 1773. In James Boswell, *The Journal of a Tour to the Hebrides, with Samuel Johnson, L.L.D., 1786.*

REFERENCES

[1]   Blum D. The blue history of cyanide – Speakeasy Science [Internet]. [cited 2017 May 6]. Available from: http://scienceblogs.com/speakeasyscience/2010/02/26/the-blue-history-of-cyanide/

[2]   Boddy-Evans M. The Accidental Creation of Prussian Blue Pigment [Internet]. [cited 2017 May 18].

Available from: https://www.thoughtco.com/artists-pigments-2573702

[3]   Starbucks Strawberry Frappuccinos dyed with crushed up cochineal bugs, report says - CBS News [Internet]. [cited 2017 May 18]. Available from: http://www.cbsnews.com/news/ starbucks-strawberry-frappuccinos-dyed-with-crushed-up-cochineal-bugs-report-says/

[4]   Prussian blue | Podcast | Chemistry World [Internet]. [cited 2017 May 18]. Available from: https://www.chemistryworld.com/podcast/prussian-blue/6101

[5]   CDC Radiation Emergencies | Facts About Prussian Blue [Internet]. [cited 2017 May 18]. Available from: https://emergency.cdc.gov/radiation/prussianblue.asp

[6]   Baskin SI, Kurche JS, Maliner BI. Physician's Guide in Terrorist Attack, Chapter on Cyanide [Internet] Humana Press , 2004 [cited 2017 May 18];263-277p. Available from: https://link.springer.com/chapter/10.1007/978-1-59259-663-8_19

[7]   Ballantyne B, Marrs TC. Clinical and experimental toxicology of cyanides. Wright 1987.

[8]   How Does Cyanide Kill? - Chemistry of Cyanide Poisoning [Internet]. [cited 2017 Apr 30]. Available from: https://www.thoughtco.com/overview-of-cyanide-poison-609287

[9]   Baskin SI, Brewer TG. Chapter 10 CYANIDE POISONING. [cited 2017 May 6]; Available from: http://www.au.af.mil/au/awc/awcgate/medaspec/ch-10electrv699.pdf

[10]   CYANIDE [Internet]. [cited 2017 May 6]. Available from: https://fas.org/nuke/guide/usa/ doctrine/army/mmcch/Cyanide.htm

[11]   Rosenberg J. The Poison Used in the Gas Chambers. [Internet]. [cited 2017 May 18]. Available from: https://www.thoughtco.com/zyklon-b-gas-chamber-poison-1779688

[12]   Gassing Victims in the Holoocaust: Zyklon-B [Internet]. [cited 2017 May 18]. Available from: http://www.jewishvirtuallibrary.org/background-and-overview-of-gassing-victims

[13]   Zyklon B. The Nizkor Project [Internet]. [cited 2017 May 18]. Available from: https://www.stormfront.org/revision/ff5zyklonb.html

[14]   Reade MC, Davies SR, Morley PT, Dennett J, Jacobs IC. Review article: Management of cyanide poisoning. Emerg Med Australas [Internet]. Blackwell Publishing Asia , 2012 Jun; [cited 2017 May 19];24(3): 225-38. Available from: http://doi.wiley.com/10.1111/j.1742-6723.2012.01538.x

[15]   Hamel J. A review of acute cyanide poisoning with a treatment update. Crit Care Nurse [Internet]. American Association of Critical Care Nurses , 2011 Feb; [cited 2017 May 19];31(1): 72-81. quiz 82. Available from: http://www.ncbi.nlm.nih.gov/pubmed/21285466 [http://dx.doi.org/10.4037/ccn2011799]

[16]   Mowry JB, Spyker DA, Brooks DE, Zimmerman A, Schauben JL. 2015 Annual Report of the American Association of Poison Control Centers' National Poison Data System (NPDS): 33rd Annual Report. Clin Toxicol [Internet] , 2016 Nov 25; [cited 2017 May 19];54(10): 924-1109. Available from: https://aapcc.s3.amazonaws.com/pdfs/annual_reports/2015_AAPCC_NPDS_Annual_Report_33rd_PD F.pdf

[17]   8 Things You May Not Know About Erwin Rommel - History in the Headlines [Internet]. [cited 2017 May 19]. Available from: http://www.history.com/news/8-things- you-may-not-know-about--rwin-rommel

[18]   "The Desert Fox" commits suicide - Oct 14, 1944 - HISTORY.com [Internet]. [cited 2017 May 19]. Available from: http://www.history.com/this-day-in-history/the-desert-fox-commits-suicide

[19]   Letter reveals Rommel's son account of his general father's last moments after being ordered to commit suicide by Hitler | Daily Mail Online [Internet]. [cited 2017 May 20]. Available from: http://www.dailymail.co.uk/news/article-2254904/Letter-reveals-Rommels-son-account-general-father s-moments-ordered-commit-suicide-Hitler.html

[20]   The usage of cyanide pills in history [Internet]. [cited 2017 Dec 5]. Available from:

https://www.thevintagenews.com/2017/05/23/the-usage-of-cyanide-pills-in-history/

[21] Adolf Hitler Biography - Biography.com [Internet]. [cited 2017 Dec 5]. Available from: https://www.biography.com/people/adolf-hitler-9340144

[22] Adolf Hitler and Ava Braun's bodyguard reveals all | Daily Mail Online [Internet]. [cited 2017 Dec 5]. Available   from: http://www.dailymail.co.uk/news/article-4334322/Bodyguard-Hitler-Eva-Braun-s-dead-bodies.html

[23] Adolf Hitler | dictator of Germany | Britannica.com [Internet]. [cited 2017 Dec 5]. Available from: https://www.britannica.com/biography/Adolf-Hitler

[24] In one German town, 1,000 people killed themselves in 72 hours [Internet] wwwTimelinecom , [cited 2017 May 19]; https://timeline.com/demmin-nazi-mass-suicide-44c6caf76727#.k9fsitd4w

[25] In one German town, 1,000 people killed themselves in 72 hours [Internet]. [cited 2017 May 20]. Available from: https://timeline.com/demmin-nazi-mass-suicide-44c6caf76727

[26] Jim Jones - - Biography.com [Internet]. [cited 2017 May 20]. Available from: http://www.biography.com/people/jim-jones-10367607

[27] Jones plotted cyanide deaths years before Jonestown - CNN.com [Internet]. [cited 2017 May 20]. Available from: http://www.cnn.com/2008/US/11/12/jonestown.cyanide/index.html

[28] CNN: Jim Jones had long planned killings - UPI.com [Internet]. [cited 2017 May 20]. Available from: http://www.upi.com/CNN-Jim-Jones-had-long-planned-killings/71471226517519/

[29] Praljak: Bosnian Croat war criminal dies after taking poison in court - BBC News [Internet]. [cited 2017 Dec 5]. Available from: http://www.bbc.com/news/world-europe-42163613

CHAPTER 6

Venoms and Poisons from the Sea

Abstract: The sea is a wondrous habitat. Humans, however, venture there at their own peril. Among the perils there are a variety of life forms that biosynthesize poisons administered *via* bites, stings and other envenomations. A few forms of sea life are poisonous when consumed by humans. Among the poisonous sea life, there are some that harbor other life forms that provide the poison. Several of the toxins described in this chapter are more annoyances than deadly; several are significant threats to human life. Sea snakes are, perhaps, the most feared of poisonous sea creatures. Sea snakes occur in the tropical and subtropical areas of the Indian and Pacific oceans ranging from the east coast of Africa to the Gulf of Panama. Most species of sea snakes, however, inhabit the waters around the Indo-Malayan Archipelago, the seas around China, and waters near Indonesia and Australia. A typical and distinguishing feature of the sea snake is the vertically flattened (paddle like) tail. They are said to be most closely related to terrestrial elapids (which include some of the most poisonous snakes, including the brown snake, taipan, death adder, cobra, krait, and mamba). Isolated but tragic instances of deaths are reported by the stingray, and food poisonings are not uncommon from several fish, due to toxins they produce or which are produced by dinoflagellates they harbor. Tetrodotoxin, produced by dinoflagellates, can be amplified up the food chain. Puffer fish poisoning is a prime example.

"And neither the angels in Heaven above Nor the demons down under the sea, Can ever dissever my soul from the soul Of the beautiful Annabel Lee"[1]

Keywords: Anemone, Blue-ringed Octopus, Box Jelly Fish, Ciguatera, Cone Shells, Crown-of-thorns Starfish, Dinoflagellates, Fire Coral, Food Chain, Lion Fish, Myoneural junctions, Paralysis, Pufferfish, Seas, Scomboid, Shell Fish, Sodium Transport Channel, Stingray, Stone Fish, Tetrodotoxin.

INTRODUCTION

A variety of sea creatures are venomous. Few people encounter them but some envenomations can be deadly. Technically, a distinction is made between poisonous and venomous. Poisonous creatures contain chemicals that are toxic when consumed. Sometimes the toxin is not made by the creature but is passed along in the food chain. Venomous creatures have poisonous chemicals in body parts that are designed to deliver the poison by bites, stings, or spines (envenomations). Both toxins and venoms will be considered. The sea snake, fire coral, lion fish, stingray, stonefish, sea anemones, blue-ringed octopus, and jelly fish deliver potent toxins by envenomation. Tetrodotoxin is an oddity; it is produced by dinoflagellates which are single-celled life forms that range from microscopic to visible in size and are in the food chain of larger sea life. Tetrodotoxin can be inflicted by the bite of the blue-ringed octopus which can accumulate the toxin, and it accumulates in the puffer fish and poisons humans who consume this fish. It is also found in certain frogs known as poison dart frogs.

SEA SNAKE

Sea snakes are probably more feared than any other poisonous creature in the sea. This reputation may be deserved because of the potency of their venom. Sea snakes occur in the tropical and subtropical areas of the Indian and Pacific oceans ranging from the east coast of Africa to the Gulf of Panama. Most species of sea snakes, however, inhabit the waters around the Indo-Malayan Archipelago, the seas around China, and waters near Indonesia and Australia [1]. There are perhaps 60 species of these marine snakes that belong to the cobra family (Elapidae) [2]. Most adults are described to be 3.3 to 5 feet long, and they have small scales that usually do not overlap but abut much like paving stones. Belly scales are small or absent and when absent, the snake "cannot crawl and is helpless on land." [2]. They give birth to young in the ocean and their typical color is alternating bands of black and gray, blue, or white rings; one variety has yellow lips [2]. The snake is usually described as not aggressive. It has a small mouth and small teeth. Stories that it cannot bite a human are, however, untrue [3]. Their fangs are short (about 0.1 inch to less than 0.2 inch). However, their mouths are flexible and it is reported that "even a small snake can bit a man's thigh… [and they] can swallow a fish that is more than twice the diameter of their neck" [3].

One source states that sea snakes are the most abundant and widespread group of poisonous reptiles in the world and estimates that there are 70 species [4]. Their venom is potent but in approximately 80% of bites, envenomation does not occur. Such bites can be inconspicuous, not cause swelling, and nearly painless. However, an adult sea snake has enough venom to kill three people [4]. With significant envenomation, death has been recorded to occur in as little as 2.5 hours. An effective antivenom has been developed and the formerly high mortality rate of approximately 10% to 50% for significant envenomation is lower but has not been accurately assessed [4].

Sea snakes are said to be most closely related to terrestrial elapids which include some of the most poisonous snakes including the brown snake, taipan, death adder, cobra, krait, and mamba [1]. The venom of the sea snake is very poisonous;

the snake is mostly shy and bites humans only when provoked, but even then it is said that they tend not to inject their venom [3]. Sea snakes breathe air and have valves over their nostrils that they close underwater. They have flattened tails used for swimming (Fig. **1**). They are usually found in shallow water and bottom-feed on fish, fish eggs, and eels. They are said to be curious creatures and are "fascinated by elongated objects such as high pressure hoses" of divers [3].

Fig. (1). The sea snake; notice the unusual flattened, paddle-like tail that is adapted for swimming.

The principal lethal factor in their venom is a neurotoxin that acts post-synaptically [5]. The neurotoxin is composed of small basic proteins consisting of 60 to 74 amino acid residues. These protein bind in a nearly irreversible way to the nicotinic acetylcholine receptors of nerve synapses and block nerve impulse transmission across myoneural junctions [5]. The venom of some species of sea snakes also contains myotoxic phospholipases. Phospholipases are enzymes that break down certain types of fats that contain phosphorous. Death results from respiratory failure caused by paralysis of respiratory muscles, primarily the diaphragm. Acute renal failure also occurs, can be fatal, and results from myoglobinuria caused by the toxin [5].

A review of the toxic properties of sea snake venom [6] describes two neurotoxins (called erabutoxins) isolated from sea snake venom. They are toxic because they block the nicotinic and acetylcholine receptors on the post synaptic membrane (Fig. **2**); thus, their action is similar to curare. This review tabulates the toxicities (in grams per nanoliter as the mouse LD_{50}) and the quantity of venom (in mL) for six species of sea snakes. The toxicities ranged from 0.38 nL to 3.5 nL (the small value indicates it is more toxic), and the quantities ranged from 0.01 ml to 0.41 mL (the larger volume is more dangerous).

My calculations of this data show that the sea snake *Astrotia stokesii* has sufficient venom to kill 5,857 mice. This snake species, of those tested, had sufficient venom to kill the most mice and this species was also the largest by a factor of approximately 5. The three-dimensional structures of the protein venoms were determined and the venoms share a common feature – three 'fingers' protruding from a dense core [6]. The presence of four disulfide bridges and two specific amino acids (out of a total of ten amino acid residues) at particular places in the peptide chain appeared to be essential for toxicity of each of the toxins examined. The size, shape and presence of particular functional groups in the molecule of toxin determine whether it will bind specifically at its site of action at the myoneural junction to interfere with nerve impulse transmission.

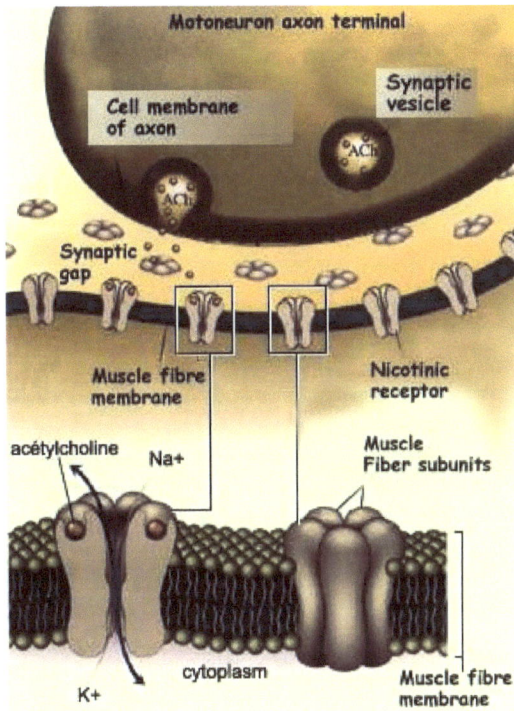

Fig. (2). Depiction of a myoneural junction, the site of action of sea snake venom. The toxin binds to and blocks the acetylcholine and nicotinic receptors; see text for details.

SEA ANEMONE

Sea anemones (Fig. **3**) and sea cucumbers are not very harmful to humans. However, a species of sea anemones, the stinging anemone (*Actinodendron plumosum*), sometimes called the Hell's Fire anemone, produces a powerful toxin with a stinger for administering it [7]. It lives in the Indo-Pacific region and burrows into the sand of the sea bottom with only its oral disk and tentacles visible. The tentacles are leaf-like or feathery and contain a toxin that is injected by stinging. There are four different genera in this species with eight different polypeptide toxins [8]. The most active of the sea anemone toxins was found to have toxicity similar to that of the most active scorpion toxins when tested on mice. The Na^+ channel of nerve cells is one site of action; however, "… scorpion toxins have affinities for the Na^+ channel structure and mechanism which are approximately 60-times higher than those found for the most active sea anemone toxins" [8]. Sea anemone also have toxins in organelles called nematocysts that assist in capturing prey. These toxins have been classified as either sodium channel toxins or potassium channel toxins [9]. Sodium channel toxins delay channel inactivation by binding to certain receptors, while potassium channel blockers bind to different receptors. These actions results in nerve paralysis and the toxins assist anemones in paralyzing prey.

Fig. (3). Sea anemone.

FIRE CORAL

Fire coral (Fig. **4**) is found in tropical and subtropical waters around the world [9]. While not a true coral (it is related to Portuguese Man-of-War), fire corals have specialized cells (cnidocytes) embedded in their calcareous skeleton. Cnidocytes contain cnidae, capsule-like structures capable of everting (turning inside-out). Cnidae that sting are called *nematocysts* (Fig. **5**). The nematocysts are used offensively and defensively and the specialized cells can eject a barbed and poisoned hook (Fig. **6**) that can paralyze their victim [10].

Fig. (4). Fire coral.

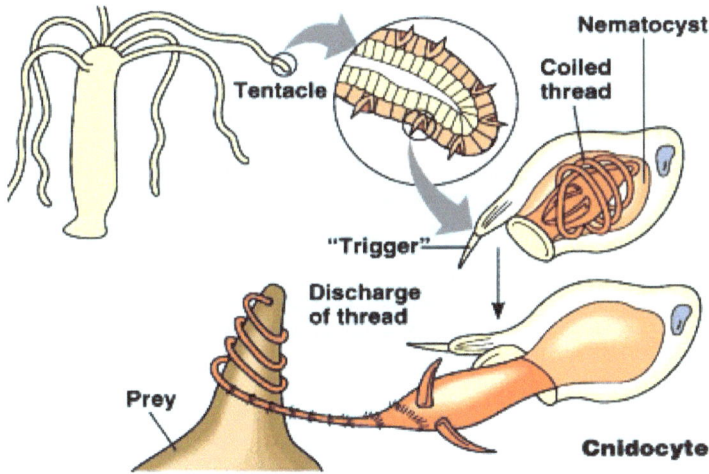

Fig. (5). Nematocysts and cnidocyte of fire coral.

Skin contact causes a mild to moderate burning sensation that lasts for hours at the poisoning site, and a rash can follow that lasts for days. When the toxin contacts an existing open wound it can cause necrosis [9]. These "stinging" corals have a hard coral-like skeleton and form large upright sheets and blades sometimes with finger-like 'antlers' and are usually yellow-green to brown in color [3].

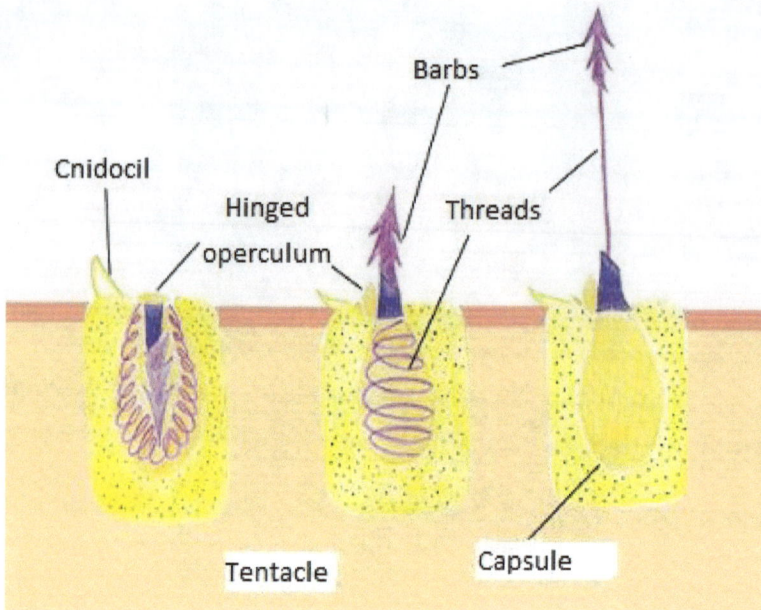

Fig. (6). Details of the mechanism of functioning of cnidae to eject barbs that can be used defensively and offensively by fire coral.

TETRODOTOXIN

Tetrodotoxin is an unusual poison in many respects. Other common names for this toxin are: fugu poison, maculotoxin, spheroidine, tarichatoxin, and TTX [11]. The Centers for Disease Control and Prevention provides basic information where it is described as an "… extremely potent poison (toxin) found mainly in the liver and sex organs (gonads) of some fish, such as the puffer fish, globefish, and toadfish… and in some amphibian, octopus, and shellfish species." Humans are poisoned when the flesh or certain organs of certain fish are improperly prepared and consumed [11]. A more complete listing of marine organisms that 'store' tetrodotoxin is found in the report "Tetrodotoxin, Molecule of the Month" for November 1999 [12]. The list includes: the Australian blue-ringed octopus, parrotfish, triggerfish, goby, angelfish, cod, boxfish (certain species), tobies, porcupine fish, molas or ocean sunfish, globefish, seastars, starfish (one species), xanthid crabs (certain species), horseshoe crab (one species), two Philippine crabs, a number of marine snails, flatworms, sea squirts, ribbonworms and arrowworms, mollusks (certain species), and marine algae. Curiously, this source also states that tetrodotoxin is found in some terrestrial organisms: the Harlequin frog (certain species), Costa Rican frog, three species of California newt, and some Salamanders; and sources say the list continues to grow.

The origin of the tetrodotoxin in species including the pufferfish was traditionally somewhat controversial, although "… it seems to be produced by endosymbiotic bacteria that often seem to be passed down the food chain" [13]. The consumption of pufferfish, which must be properly prepared by a skilled chef to be safe, is "considered the most delicious fish in Japan" [13]. There is no known antidote to tetrodotoxin which inhibits sodium channels in cells and can produce heart failure. In Japan, a regulatory limit of 2 mg of tetrodotoxin per kilogram of fish for consumption exists [13]. Tetrodotoxin is a powerful neurotoxin and is approximately 1,200 times as toxic to humans as is cyanide [13].

Toxnet [14] provides human health effects described *via* excerpts of poisoning accounts and signs and symptoms are included. When a toxic amount of tetrodotoxin is consumed, the most significant effects are neurological and gastrointestinal toxicity. With severe poisoning, there may be dysrhythmias and hypotension, and death can result. The Toxnet site (of the National Library of Medicine) [14] states that the clinical description of tetrodotoxin poisoning is met if there is rapid onset of one of the following neurological and gastrointestinal signs and symptoms: "1) oral paresthesia (might progress to include the arms and legs, 2) cranial nerve dysfunction, 3) weakness (might progress to paralysis), or 4) nausea or vomiting." One may presume that these symptoms most likely occur in a restaurant setting after or during consumption of suspicious food. It has a violent onset that is hard not to notice by all those present.

A study described 100 cases of human intoxication by tetrodotoxin in Taiwan from 1998 to 2008 (references were cited in the original) and was reported by Lago and co-workers [13]. One incident affected eight people with five deaths. In another reported incident, 37 people became ill with manifestations of pufferfish poisoning. "Twenty-two patients developed ascending paralysis, weakness of both limbs, and the respiratory muscle (sic) were involved in other patients. Fourteen patients had manifestations within 30 minutes of ingestion. Out of these 37 cases, eight patients died within five hours of (sic) post-ingestion. The cause of death in all these patients was respiratory muscle paralysis leading to respiratory failure." This site [13] provides a table grading tetrodotoxin intoxication based on symptoms. The table should be consulted for explicitness; however, four grades were described. The lowest grade (grade 1), showed neuromuscular symptoms (paresthesia around the mouth, headache, diaphoresis, papillary constriction) and mild gastrointestinal symptoms. Grade two was defined as parenthesis spreading to the trunk and extremities, early motor paralysis and lack of coordination. Grade three was defined as neuromuscular symptoms that tend to increase (dysarthria, cysphagia aphagia, lethargy, incoordination, ataxia, floating sensation, cranial nerve palsies, muscular fasciculations) cardiovascular/pulmonary symptoms [see source for details]; and dermatologic symptoms [see source for details]. Stage four was defined as: "Impaired conscious state, respiratory paralysis, severe hypotension, and cardiac arrhythmia."

The flesh of many pufferfish, properly prepared by a qualified chef, may not actually be dangerous to consume; however, the gonads, liver, intestines, and skin of pufferfish can contain lethal amounts of tetrodotoxin [14]. Tetrodotoxin has also been found in many species including: "… the California newt, parrotfish, frogs of the genus *Atelopus*, the blue-ringed octopus, starfish, angel fish, and xanthid crabs" [14]. It is reported that tetrodotoxin is responsible for 30 to 50 cases of intoxications each year [13].

The principal cite and action of tetrodotoxin (Fig. **7**) is to block voltage-gated sodium channels (Fig. **8**) to cause paralysis [13]. This reduces membrane excitability of vital cells, tissues and organs including heart myocytes, skeletal muscles, and the central and peripheral nervous systems. This information is provided by Lago and associates with referenced sources [13]. The median, lethal dose (LD_{50}) of tetrodotoxin for mice for intragastric administration was 532 µg per kilogram of mouse and much less by intraperitoneal or subcutaneous injection (10.7 and 12 µg per kilogram, respectively). The minimum lethal dose was even lower in rabbits (about half that of mice, as stated above).

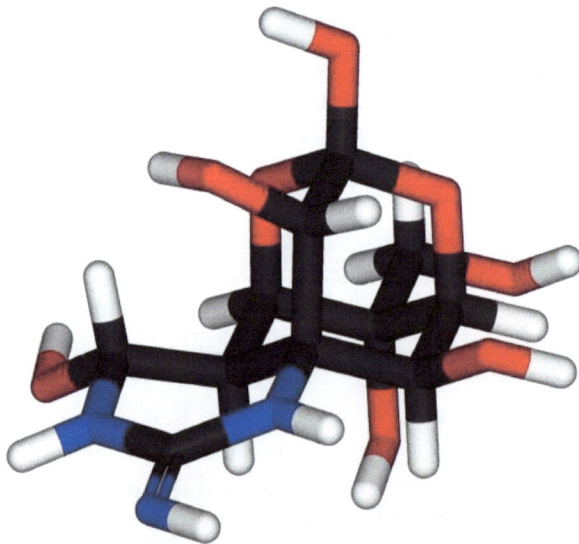

Fig. (7). Model of a molecule of Tetrodotoxin. Black represents carbon and its bond; red is oxygen and its bond, blue is nitrogen and its bond, and white is hydrogen and its bond.

Tetrodotoxin mimics the hydrated sodium cation and enters the mouth of the sodium channel [12]. The arrangement of amino acid in the membrane channel favors the attraction of the tetrodotoxin and subsequently the tetrodotoxin molecule becomes electrostatically attached to the opening of the sodium channel. While hydrated sodium binds reversibly on the nanosecond time-scale, tetrodotoxin remains on the site on the order of tens of seconds [12]. "With the bulk of the [tetrodotoxin] molecule denying sodium the opportunity to enter the channel, sodium movement is effectively shut down, and the action potential along the nerve membrane ceases" [12]. This site also states: "A single milligram or less of [tetrodotoxin] – an amount that can be placed on the head of a pin, is enough to kill an adult [human]."

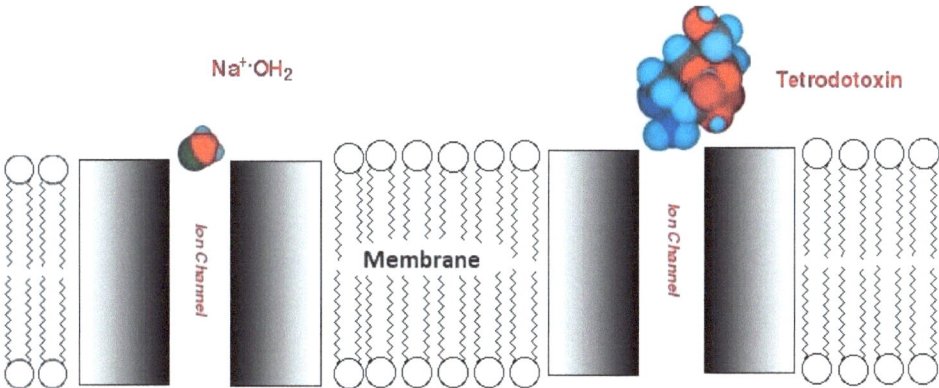

Fig. (8). Tetrodotoxin blocks sodium ion channels in membranes and its principal mechanism of toxicity is to block nerve impulses.

Historically, tetrodotoxin was first isolated by Yoshizumi Tahara in 1894 from the ovaries of the globefish, and in 1909 he showed that from this source there was only one toxic substance and he named it tetrodotoxin. This word is derived from the name of the family of pufferfish from which he isolated the poison [13]. This source further states, that for some time, it was believed that tetrodotoxin was present exclusively in the pufferfish family and it was controversial regarding whether the toxin was produced by the pufferfish or was taken in from the environment and accumulated in the fish. This source also states that tetrodotoxin was found in the California newt in 1964 and subsequently has been found in other marine animals. Pufferfish are non-toxic when artificially reared with diets not containing tetrodotoxin. However, they become toxic with accumulated tetrodotoxin when the diet is changed to contain tetrodotoxin. This confirms that the food chain is the source of the toxin. Marine bacteria are said to be the primary source of the tetrodotoxin [13].

Fugu (pufferfish) poisonings totaled 646 reported cases from 1974 through 1983 from Japan alone, and there were 179 fatalities [12]. This site also states:

"Estimates as high as 200 cases per year with mortality approaching 50% have been reported. Only a few cases have been reported in the United States... Sushi chefs... in Japan... must be licensed by the Japanese government".

The comparative toxicity of tetrodotoxin was summarized by Jim Johnson [12] who cited a reference by William Light: "[W]eight-for-weight, tetrodotoxin is ten times as deadly as the venom of the many-banded krait of Southeast Asia. It is 10 to 100 times as lethal as black widow spider venom (depending upon the species) when administered to mice, and more than 10,000 times deadlier than cyanide." The following is also taken from this source. "It has the same toxicity as saxitoxin which causes paralytic shellfish poisoning... [which also blocks the sodium channel]... and both are found in the tissues of pufferfish... Except for a few bacterial proteins, only palytoxin, a bizarre molecule isolated from certain zoanthideans [which resemble sea anemones]... and mailotoxin... found in certain fishes associated with ciguatera poisoning, are known to be significantly more toxic than [tetrodotoxin]. Palytoxin and mailotoxin have potencies nearly 100 times that of [tetrodotoxin] and saxitoxin, and all four toxins are unusual in being non-proteins... there is also some evidence for a bacterial biogenesis of saxitoxin, palytoxin, and mailtotoxin... [in] living animals the toxin acts primarily on myelinated (sheathed) peripheral nerves and does not appear to cross the blood-brain barrier" [12].

As far as I have been able to determine, little is known about the specific mechanisms of biosynthesis of tetrodotoxin. This is of interest because of the large number of sea creatures that 'store' the toxin in their tissues and the fact that it is also found in a few terrestrial life forms, none of which appear to biosynthesize the toxin. As cited by Johnson [12], enzymes capable of biosynthesizing tetrodotoxin have not been isolated from any source. There is, however, strong evidence in research (from the U.S. Food and Drug Administration and the National Food Agency of Denmark) cited by Johnson [12], of a bacterial origin for tetrodotoxin. Summarized, this evidence includes: pufferfish grown in culture do not produce tetrodotoxin until they are fed tissues from a toxin producing fish; the Australian blue-ringed octopus accumulates tetrodotoxin in a special salivary gland and injects its prey with toxin by bite, and the octopus contain tetrodotoxin-producing bacteria; and certain xanthid crabs contain both tetrodotoxin and paralytic shellfish toxin. Johnson also states: "It is now clear that marine bacteria have long been in mutualistic symbiosis with marine animals... It is now known that the related toxins tetrodotoxin and anhydrotetrodotoxin are synthesized by several bacterial species, including strains of the family *Vibrionaceae, q.v., Pseudomonas sp.,* and *Photobacterium phosphoreum.* Following grazing, marine invertebrates and vertebrates accumulate these bacteria, provide them with a suitable host environment, and in return receive the protection of marine biotoxins, compliments of the prokaryotes" [12]. I have quoted extensively from these sources to provide the information as accurately as possible. It should be noted that this evidence is in addition to the evidence for dinoflagellates as a source which is provided in the following paragraphs.

Blue-Ringed Octopus, a Source of Tetrodotoxin Poisoning

When I first learned about the blue-ringed octopus, I envisioned a beautiful creature that possibly grew to monstrous size. I have subsequently learned that it certainly is beautiful, but it truly is shy and retiring and it is tiny in size (Fig. **9**). There are many species of this creature and they are truly small and range from about one and one-half inches to two and one-third inches long, with tentacles approximately two and three-quarters inches to four inches long [15]. The blue circles are seen only when the animal is threatened and apparently serve as a warning of the danger they pose. The blue color is a pigment in chromatophores in the eight legs (or arms) that are arranged around the mouth. The arms contain suckers as is typical for all octopi. The creature has three hearts, very large eyes, and gills for breathing water [15]. Water can be forcefully ejected to propel the octopus rapidly, and there are glands that make ink that can be ejected to hide the animal. Curiously, the blood of the octopus is a transparent blue color which is due to the oxygen-carrying component of the blood which contains copper. The respiratory pigment of the octopus is similar to hemoglobin which is red and contains iron, not copper. Octopi technically are mollusks, which includes the quite dissimilar snails, slugs, and bivalves. The range of the blue-ringed octopus is from the Sea of Japan to the seas of southern Australia [15].

The toxin of the blue-ringed octopus is in its saliva and is called tetrodotoxin [12, 15]. This word, apparently, is derived from Latin words for four, tooth, and poison. Various accounts report about deaths from bites by the blue-ringed octopus. For example, one account states that there have been at least 3 documented deaths (two in Australia and one in Singapore) [15]. Dr. Ray Caldwell, a biologist who has had an encounter with the creature, reports knowledge of the death of a German citizen found dead on an Andaman Sea beach, and also states: "… several humans suffer bites each year… The bites are slight and produce at most only a small laceration with no more than a tiny drop of blood and little or no discoloration. Bites are usually reported as painless" [12]. Caldwell also says there is "… some question as to whether the octopus even needs to envenomate a human [to produce intoxication]… I can report developing mild local neurological symptoms after immersing my hand in sea water in which a large blue-ringed octopus had been shipped." Caldwell also describes symptoms from being poisoned [12]:

"Depending on how much venom has been transferred into the wound, the onset of symptoms can be quite rapid. Within five to ten minutes the victim begins to experience parasthesis and numbness, progressive muscular weakness and difficulty breathing and swallowing. Nausea and vomiting, visual disturbances and difficulty speaking may also occur. In severe cases, this is followed by flaccid paralysis and respiratory failure, leading to unconsciousness and death due to cerebral anoxia…the victim's heart continues to beat until extreme asphyxiation sets in. Some victims report being conscious, but unable to speak or move. They may even appear clinically dead with pupils fixed and dilated."

Fig. (9). Blue-ringed octopus showing the iridescent circular chromatophores which appear when the animal is disturbed and are characteristic of this genus that contains many species.

Puffer Fish, a Source of Tetrodotoxin Poisoning

The puffer fish (Fig. **10**) can poisons humans who consume them. This creature is also known as: blowfish, swellfish, globefish, balloonfish, and bubblefish [12]. This source describes them as slow, clumsy swimmers and rather than trying to escape predators, they inflate with water and air and form a ball shape that is difficult to ingest by predators. Some species also have spines as defenses. The scientific name for this group (*Tetradon*) refers to the fact that they have four large teeth which they use to crush their prey which is crustaceans and mollusks [12]. This source also states that "The eyes and internal organs of most pufferfish are highly toxic, but nevertheless, the meat is considered to be a delicacy in Japan and Korea." The pufferfish is toxic because of tetrodotoxin stored in their tissues [12].

Fig. (10). Pufferfish, with spines and inflated with air and water to make it difficult for a predator to swallow.

CIGUATERA DINOFLAGELLATES AND TOXINS

Ciguatera is a food borne illness transmitted by eating certain reef fish contaminated with a toxin synthesized by some dinoflagellates such as *Gambierdiscus toxicus* that live in tropical and subtropical waters [16]. These small single-celled organisms adhere to plant material consumed by herbivorous fish which in turn are consumed by larger carnivorous fish such as barracuda, shark and also by omnivorous fish including bass and mullet. This process results in biomagnification (increased concentration of the toxin) from species low on the food chain to organisms higher on the food chain. Affected fish may show no signs of illness but can cause ciguatera in humans who consume them.

Dinoflagellates (Fig. **11**) are sometimes referred to as protists (a eukaryote that is not classified as a plant, animal, or fungus). They are single-celled organisms with two flagella. They are part of marine plankton but also are found in fresh water. Some synthesize toxins that accumulate in marine species including fish and shellfish. The word dinoflagellate means 'possessing two whirling flagella' by rough translation. Flagella are whip-like organs of motion and dinoflagellates usually possess two. Because dinoflagellates biosynthesize poisons that affect humans in several ways, they are considered here as a topic with subcategories for several poisoning source from sea life.

Dinoflagellates are a very diverse group. Some are microscopic but the largest can reach 2 millimeters in diameter. Many are photosynthetic, some are bioluminescent, and some are parasites of fish and shellfish. Dinoflagellates flourish is some coastal waters and during the warm season they can multiply intensely to create red or golden water called 'red tide'. The red tide dinoflagellates produce a neurotoxin and a significant fish kill can result that may be ecologically significant and disturbing to people in the area.

In recent times, concern has been expressed by some environmentalists about the possibility of toxic effects for humans living near such coasts because of airborne toxins. Humans who consume parasitized fish can become ill with a disease called ciguatera, or can be afflicted with paralysis after consuming clams, mussels, or oysters parasitized with dinoflagellates. These human diseases are serious but not usually fatal [17].

This dinoflagellate produces several, similar toxins including ciguatoxin, maitotoxin, gambieric acid, and scaritoxin [18]. These toxins are polyethers and a long-chain alcohol (palytoxin) and are not destroyed, as are protein toxins, by cooking the fish. Ciguatoxin is said to be lethal at 0.45 µg/kg when injected intraperitoneally in mice; and maitoxin is lethal at a dose of 0.15 µg/kg. Oral intake of as little as 0.1 µg of ciguatoxin can cause illness in an adult human [18]. Ciguatoxin opens sodium channels in nerve cell membranes and in striated

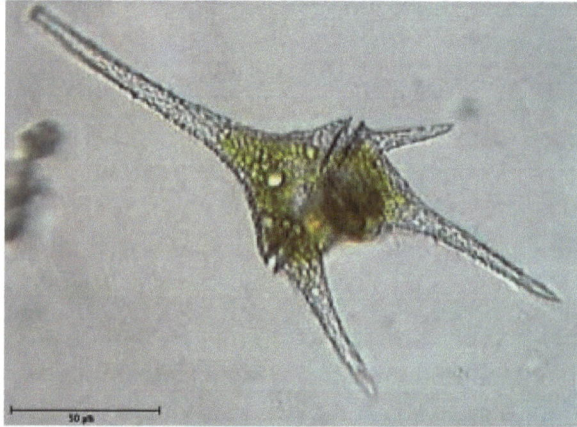

Fig. (11). Microscopic view of a dinoflagellate, a single-celled organism. The size indicator on the bottom left is 50 microns (bacteria are typically one or a few microns in diameter or length, but may be several microns). This cell apparently possesses chlorophyll (green coloration) and is therefore photosynthetic and not parasitic.

muscle in the heart. Palytoxin causes severe contraction of all types of muscles and it is a skin irritant. Symptoms of poisoning are long-lasting and damage to nerves is only repaired by regeneration of nerves. Ciguatoxin causes respiratory arrest *via* depression of the respiratory center [18]. Ciguatera is said to be "… the most commonly reported marine toxin disease in the world" [18].

Relative to the 'art of poisons', broadly interpreted in history, ciguatera may have been the reason for some historically famous Polynesian voyages [17]. Cultural exchange and colonization was said to be particularly active from AD 100 to 1450. A question asked by the authors of a guest editorial in the Journal of Biogeography in 2009 [17] is: "… why would large groups of people leave their homelands to voyage into the unknown? Oceanic voyages are risky" These authors hypothesized that ciguatera prompted voyages in this era, and sought answers from three areas of research: archaeological evidence, ciguatera fish poisoning reports since the 1940s, and climate and temperature oscillations from what they called 'palaeodatasets'. They showed "… that ciguatera fish poisoning events coincide with Pacific Decadal Oscillations and suggest that the celebrated Polynesian voyages across the Pacific Ocean may not have been random episodes of discovery to colonize new lands, but rather voyages of necessity. A modern analogue (in the 1990s) is a shift towards processed foods in the Cook Islands during ciguatera fish poisoning events, and mass migration of islanders to New Zealand and Australia."

It is evident that poisons and poisonings can be dramatic and tragic for individuals. Indeed, ciguatera may be even more – an example of an unusual type of poisoning drastically affecting a whole nation of people as a driving force in the history of the 'migration' of a people with even wider implications for paleogeology.

STONE FISH

Stonefish resemble their name (Fig. **12**). They are commonly found in shallow waters of the Pacific and Indian Oceans [19, 20]. They are described to have a large head and skin covered with lumps that can easily blend in with their environment [20]. The venom, a proteinaceous toxin called verrucotoxin, is stored in the dorsal fin spines [19]. These spines are grooved and envenomation of a human usually occurs when an unaware person steps on the fish [20]. The fish commonly grows to about 13 inches in length and they are said to be sluggish bottom-dwellers, living among the coral and also in mud-flat estuaries [20].

The injected verrucotoxin induces intense pain, respiratory arrest, damage to the cardiovascular system, convulsion and skeletal muscle paralysis which can lead to death [19]. The toxin is a glycoprotein with a molecular weight of approximately 332,000 Daltons consisting of two proteins of approximately 166,000 Daltons each [19]. The toxin modulates calcium^{2+} activity through a specific receptor pathway. In the heart, it is postulated that the toxin produces arrhythmia and may also activate certain proteases [19].

Fig. (12). The stonefish is sometimes camouflaged in coral or mud, and when stepped upon by a person, a toxin is injected through dorsal spines.

BOX JELLY FISH

The box jellyfish is said not to be a 'true' jellyfish and it is classified differently but has many of the same characteristics as a true jellyfish [21]. The main body of the animal (the bell) appears to have four sides and hence its name 'box' (Fig. **13**). The most deadly species, and also the largest, is the Australian box jellyfish (*Chironex fleckeri*). However, *Carukia barnesi* also has caused human deaths and is tiny, only about as big as one's thumbnail [21]. The Australian box jellyfish when fully grown can be up to approximately 8 inches measured along each box side with tentacles that are each up to 10 feet in length [22]. They can weigh as much as approximately 4 pounds. There are approximately 15 tentacles on each corner of the 'box' and tentacles have thousands of specialized stinging cells called nematocysts which can be activated by contact with the surface of fish, shellfish or (most unfortunately) humans. Jellyfish nematocysts can sting even when the tentacle containing the nematocysts are severed from the body [21, 23]. Their transparency and blue color hides them from view in the sea. The box jellyfish move by a jet-like mechanism and can reach more than 3.5 to 4 miles per hour – approximately up to 6.5 ft per second. In contrast, most jellyfish just float. The box jellyfish posses sophisticated eyes [21].

The Australian box jellyfish's habitat is the tropical oceans around northern Australia, and it is also present in the waters of the Indo-Pacific region near Papua New Guinea, the Philippines and Vietnam [21]. It is said that they favor river mouths and estuaries and creeks, especially after rains and move into shallow waters during rising tides. They do not like deep water, rough seas, coral reefs or areas with sea grass [21].

They have venom that is far different from the 'true' jellyfishes. Because of its delicate nature, it is easy to speculate that the box jellyfish needed a very quick-acting venom to very quickly incapacitate any sea animal that it sought as prey. The venom, present in the nematocysts previously mentioned, causes several toxic effects on the skin, the heart, and the nerves of humans [21, 23]. Victims with extensive contact with tentacles can experience cardiac arrest within minutes [21].

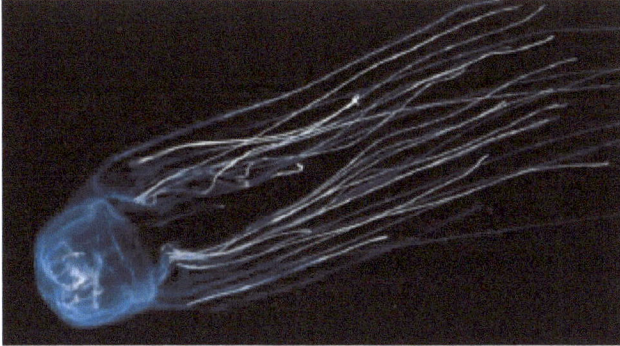

Fig. (13). Box jellyfish.

An account of a 20-year-old German woman who was a tourist in Thailand and while swimming became entangled in the tentacles of a box jellyfish and was killed may be typical [24]. A friend was stung on a hand and survived. This account states that the death was the third reported fatality in Thailand over the previous 14 months from poisoning by the box jellyfish.

A review of jellyfish toxins in 2015 [25] states: "Although countless numbers of people are being stung every year by poisonous jellyfish throughout the world, statistical data is rarely available, even for fatal cases, except in Australia... only a few toxic components... have been identified... probably because it is difficult to collect jellyfish venom of high purity... venoms [identified to date] are proteins ... intrinsically susceptible to... loss of biological activity during extraction and purification procedures [25]. This appears to account for the lack of information in the literature about specific toxins, none of which have received names. A review that I conducted using Ovid Medline, covering 1946 to the present, identified only 26 papers focused on identification of box jellyfish toxins. Typical of the most instructive of these was an article describing box jellyfish protein venoms [26]. This article states: "The box jellyfish *Chironex fleckeri* produces extremely potent and rapid-acting venom that is harmful and lethal to prey." They isolated two venom proteins, CfTX-A and CfTX-B, that were large molecules: ~ 40,000 Da and ~ 42,000 Da, respectively, in size (Da refers to a unit used to quantify the mass of a large molecule such as a protein) [26]. Using bioinformatics analysis involving use of cDNA library sequences, they identified the sources of these toxins plus two other toxins they named: CfTX-1 and CfTX-

2. For simplicity I shall refer to these toxins as 1, 2, A, and B. Comparative bioactivity assays showed that 1 and 2 caused profound effects on the cardiovascular system of an anesthetized rat; whereas, A and B elicited only minor effects at the same doses (25 nanograms of toxin per kilogram of rat). Conversely, the hemolytic activity of A and B toxins were at least 30-times as great as that of 1 and 2.

A report published in 2015 [27] gives further information about incidences and prevalence of poisoning by the box jellyfish. It states that a toxic jellyfish network and its surveillance system were set up, respectively, in 2008 and 2009. Three cases with severe cutaneous injuries were described in this article which also provides some history of recorded incidences. It states that there were 57 probable cases of box jellyfish stings between 2008 and 2013. It further characterizes that "… toxins produced by these animals are a complex mixture of polypeptides and proteins, with hemolytic, cardio toxic, and dermonecrotic effects" (with references provided). The box jelly fish poisoned site shows the effect of the impact of the whip-like appendage on the victim's skin and is purple to brown in color. Wounds become erythematic, edematous with vesicle formation and may progress to full-thickness necrosis within the first two weeks following the sting. Long-term complications of these wounds include keloids, granulomas, hyper- pigmentation, fat atrophy, and muscle contractures [27].

CONE SHELLS

Cone shell poisoning is rare but is included here for completeness. This animal includes 500 species [28]. They have very colorful, attractive shells that are sold commercially (Fig. **14**). All are carnivores and feed on other mollusks, worms, and about 70 species are known to feed on small fish. They paralyze their prey by injecting poison using a harpoon-like appendage connected to a muscular poison gland [28]. It is said that the cone shell can selectively alter the toxin composition and amount injected to conserve toxin and to adjust the envenomation to fit the prey. The venom can be sufficient to kill a human [28].

The venom apparatus is intricate. Venom is applied using teeth called radula (translation: rasp tongue); however, in the cone shell, there typically are only a small number of teeth [29]. The venom gland is tube-shaped [28]. The toxins are oligopeptides (chains of amino acids, shorter in length than proteins). The animal is designed to prevent self-poisoning. This is achieved by biosynthesis of the toxins from non-poisonous intermediates that are stored. Shortly before injection, the non-poisonous molecules are chemically united into the effective nerve poison known as conotoxin [28, 30].

The venom is known to be lethal for humans. It is composed of neurotoxins called conotoxins. All have low molecular weights and include components that are extremely fast-acting, which benefits the slow-moving cone shell (snail) [29]. The toxin has two components that differ in time required for toxic effect. Both affect (block) neuromuscular junctions. One component causes immediate immobilization (designated as a lightning strike) by peptides that inhibit voltage-gated sodium channels and other peptides that block potassium channels. This causes a massive depolarization of all axons (nerves) in the near vicinity of the venom injection site. This effect is analogized to be similar to electrocution of the prey. The second effect is slower and involves the total inhibition of neuromuscular transmission which is effected by chemicals called conopeptides. These toxins act at sites remote from the venom injection site and the effect occurs specifically at myoneural junctions. Most specifically, the sites of action are multiple: the presynaptic calcium channels that control neurotransmitter release, the postsynaptic nicotinic receptor sites, and the sodium channels that control muscle action potential (to effect muscle contraction) [29].

Fig. (14). Cone shells; all have the characteristic cone shape and often are brightly colored. They are sold commercially.

The number of deaths from cone shell envenomation can only be approximated. Fifty death are commonly cited to have been described and the death rate appears to be approximately 25% when there is serious envenomation [29]. Typically, there is intense burning at the sting site and this effect leads in about one hour to progressive paralysis of body muscles without other signs and symptoms. In later stages there can be blurred vision and difficulty in speaking, and unconsciousness with difficulty in breathing and possible respiratory arrest that can be fatal in 40 minutes to 5 hours after the sting [29].

The action of cone shells (cone snails) has been summarized as: "… a cocktail of polypeptides that mainly target different voltage- and ligand-gaited ion channels. Typically, conopeptides consist of 10 to 30 amino acids but conopeptides with more than 60 amino acids have also been described" [31]. This paper, published in 2008, also describes the biotechnological production of these peptides for research and medical applications that are on-going.

Crown-of-Thorns Starfish, Lion Fish, Stingray, and Scomboid Toxin

I will combine discussion of these three marine creatures since they are toxic, but relatively low in potency and significance for poisoning humans.

The crown-of thorns starfish (*Acathaster planci)*, so-named because of its appearance as similar to the crown placed on Christ's head when He was crucified, has spines that contain venom (Fig. **15**). This marine species has gained a bad reputation from being involved in coral reef destruction. Skin injuries to humans occur and are considered medically serious [32]. Stings from this organism produce pain, protracted vomiting, and erythema (redness) and swelling at the envenomation site. Crude toxin has various biological activity including: hemolysis, lethality for the mouse, edema formation, phospholipase activity, and anticoagulant activity [32]. Two lethal components have been identified from crown-of-thorn starfish: plancitoxins I and II, known as major and minor toxins, respectively [33]. However, these toxins had equivalent toxicities, in terms of the LD_{50} when injected into mice, of 140 µgram per kilogram of mice [33]. This is equivalent to 2.8 µg of venom to kill one mouse on average. The authors [33] also reported that the venom was toxic to the liver. The cytotoxicity of venom was investigated using five cell culture lines including cancer cell lines and normal human cell lines. Toxicity was documented for all cell lines and the results were interpreted to suggest that the venom had potential for chemotherapy [34].

Fig. (15). Crown of thorns.

Lion fish (*Ptervis* sp.) are native to the Indo-Pacific Oceans, but they have spread to warm regions of other marine environments, particularly the western Atlantic, and this is of concern. They have venomous spines containing a poison but they are not considered deadly to a healthy person (Fig. **16**). The venom is a combination of a protein, a neuromuscular toxin, and the neurotransmitter acetylcholine. They are of potential concern as a source of ciguatoxins and have recently spread to the Caribbean, including the French Antilles which is a ciguatera-endemic region. They are consumed in some cultures and are a popular aquarium species. Curiously, they have been described to be stridulate (sound producing) species [35 - 37].

Fig. (16). Lion fish.

The stingray is included here because it is clinically import in the United States where it accounts for an estimated 1,500 injuries annually; however, most of these are minor. Smaller specimens can partially bury themselves in shallow water and people may step on them. They have a long and powerful tail with a spine containing a venom gland. They grow quite large and can lash out quickly and repeatedly with the tail which they direct forward over their backs. Sometimes the venomous spine breaks and remains in the wound. The venom can lead to local tissue cyanosis and necrosis and may become infected or otherwise heal poorly [23]. The venom contains the neurotransmitter serotonin which causes severe pain at the sting site, and two enzymes: 5-nucleotidase and phosphodiesterase which cause tissue necrosis. Symptoms of envenomation are immediate and intense pain, salivation, nausea, vomiting, diarrhea, muscle cramps, dyspnea, seizures, headache, and cardiac arrhythmias. Most stings, especially by smaller animals, are not clinically serious. However, large stingrays with large powerful tails and long spines can cause exsanguinations when they penetrate a vital organ. Death can result from this rather than from the effects of the poison [23].

A tragic death illustrates an extreme attack by a stingray that may have weighed as much as 200 pounds (Fig. **17**). Steve Irwin, famously known as 'The Crocodile Hunter' had many dangerous encounters with wild animals of all kinds including poisonous reptiles. He was killed by a stingray but he did not die from the poison but primarily from exsanguination and heart damage from having his chest repeatedly punctured by the spine of a stingray [38].

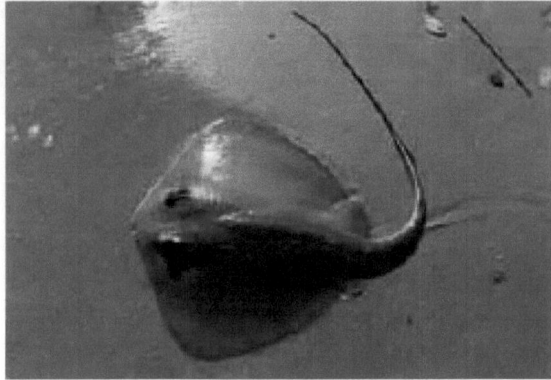

Fig. (17). Stingray.

Scomboid will be briefly mentioned. It results from improper handling of fish between the time it is caught and when it is cooked and eaten. Improper preservation and refrigeration can lead to production of histamine and histamine-like chemicals derived from the dark meat of certain fishes including tuna and mackerel. The effects develop rapidly, within 20 to 30 minutes, after eating spoiled fish. Symptoms, which typically may last less than six to eight hours, include any or all of the following: flushing, nausea, vomiting, diarrhea, severe headache, palpitations, abdominal cramping, dizziness, dry mouth, a rash of red welts on the skin that itch intensely, and congested blood vessels in the eyes [23].

NOTES

[1] Edgar Allen Poe (1808-1849). *Annabel Lee [1849], st. 5*

REFERENCES

[1]   Rasmussen AR. SEA SNAKES. [cited 2017 Nov 13]; Available from: http://www.fao.org/ 3/a-y0870e/y0870e65.pdf

[2]   sea snake | reptile | Britannica.com [Internet]. [cited 2017 Nov 13]. Available from: https://www.britannica.com/animal/sea-snake

[3]   Fire Coral | Hazardous Marine Life - DAN Health & Diving [Internet]. [cited 2017 Nov 13]. Available from: https://www.diversalertnetwork.org/health/hazardous-marine-life/fire-coral

[4]   Sea Snake Envenomation: Background, Pathophysiology, Epidemiology [Internet]. [cited 2017 Nov

13]. Available from: https://emedicine.medscape.com/article/771804-overview#a4

[5]   Takasaki C. The Toxinology of Sea Snake Venoms. J Toxicol Toxin Rev [Internet] Taylor & Francis , 1998 Jan 2; [cited 2017 Nov 13];17(3): 361-72. [cited 2017 Nov 13] http://www.tandfonline.com/ doi/full/10.3109/15569549809040398

[6]   Tamiya N, Yagi T. Review Studies on sea snake venom. [cited 2017 Nov 13]; Available from: https://www-jstage-jst-go-jp.proxy.mul.missouri.edu/article/pjab/87/3/87_3_41/_pdf

[7]   Venomous Species [Internet]. [cited 2017 Jan 1]. Available from: http://www.redang.org/fishdanger. htm

[8]   Schweitz H, Vincent JP, Barhanin J, Frelin C, Linden G, Hugues M, *et al.* Purification and pharmacological properties of eight sea anemone toxins from Anemonia sulcata, Anthopleura xanthogrammica, Stoichactis giganteus, and Actinodendron plumosum. Biochemistry [Internet] , 1981 Sep 1; [cited 2017 Nov 13];20(18): 5245-. Available from:http://www.ncbi.nlm.nih.gov/ pubmed/6117312

[9]   Honma T, Shiomi K. Peptide Toxins in Sea Anemones: Structural and Functional Aspects. Mar Biotechnol [Internet] Springer-Verlag , 2006 Jan 1; [cited 2017 Nov 13];8(1): 1-10. Available from: http://link.springer.com/10.1007/s10126-005-5093-2

[10]   Brusca RC, Brusca GJ. Invertebrates. Sinauer Associates 2003.

[11]   CDC - The Emergency Response Safety and Health Database: Biotoxin: TETRODOTOXIN - NIOSH [Internet]. [cited 2017 Nov 14]. Available from: https://www.cdc.gov/niosh/ershdb/ emergencyresponsecard_29750019.html

[12]   What makes blue-rings so deadly? - The Cephalopod Page [Internet]. [cited 2017 Nov 15]. Available from: http://www.thecephalopodpage.org/bluering2.php

[13]   Lago J, Rodríguez LP, Blanco L, Vieites JM, Cabado AG. Tetrodotoxin, an Extremely Potent Marine Neurotoxin: Distribution, Toxicity, Origin and Therapeutical Uses. Mar Drugs [Internet] Multidisciplinary Digital Publishing Institute (MDPI) , 2015 Oct 19; [cited 2017 Nov 14];13(10): 6384-406. Available from:http://www.ncbi.nlm.nih.gov/pubmed/26492253

[14]   TETRODOTOXIN - National Library of Medicine HSDB Database [Internet]. [cited 2017 Nov 14]. Available from: https://toxnet.nlm.nih.gov/cgi-bin/sis/search/a?dbs+hsdb:@term+@DOCNO+3543

[15]   Blue ringed octopus - AIMS [Internet]. [cited 2017 Nov 15]. Available from: https://www.aims. gov.au/docs/projectnet/blue-ringed-octopus.html

[16]   Identifying Harmful Marine Dinoflagellates - Gambierdiscus toxicus / Department of Botany, National Museum of Natural History, Smithsonian Institution [Internet]. [cited 2017 Nov 16]. Available from: http://botany.si.edu/references/dinoflag/Taxa/Gtoxicus.htm

[17]   Introduction to the Dinoflagellata [Internet]. [cited 2017 Nov 14]. Available from: http://www. ucmp.berkeley.edu/protista/dinoflagellata.html

[18]   Ciguatera Fish Poisoning : Red Tide [Internet]. [cited 2017 Nov 16]. Available from: http://www. whoi.edu/redtide/page.do?pid=9679&tid=523&cid=27687

[19]   Yazawa K, Wang J-W, Hao L-Y, Onoue Y, Kameyama M. Verrucotoxin, a stonefish venom, modulates calcium channel activity in guinea-pig ventricular myocytes. Br J Pharmacol [Internet] Wiley-Blackwell , 2007 Aug; [cited 2017 Nov 15];151(8): 1198-203. Available from:http://www. ncbi.nlm.nih.gov/pubmed/17572694
        [http://dx.doi.org/10.1038/sj.bjp.0707340]

[20]   stonefish | fish, Synanceiidae family | Britannica.com [Internet]. [cited 2017 Nov 15]. Available from: https://www.britannica.com/animal/stonefish-Synanceiidae-family

[21]   The Australian Box Jellyfish [Internet]. [cited 2017 Nov 15]. Available from: http://www.outback-australia-travel-secrets.com/box-jellyfish.html

[22] US Department of Commerce NO and AA. What is the most venomous marine animal? [cited 2017 Nov 15]; Available from: https://oceanservice.noaa.gov/facts/box-jellyfish.html

[23] Perkins RA, Morgan SS. Poisoning, envenomation, and trauma from marine creatures. Am Fam Physician 2004; 69(4): 885-90.
[PMID: 14989575]

[24] German tourist killed by a box jellyfish at popular Thai resort | Daily Mail Online [Internet]. [cited 2017 Nov 15]. Available from: http://www.dailymail.co.uk/travel/travel_news/article-3264557/Pictured-20-year-old-tourist-killed-box-jellyfish-popular-Thai-resort-entangled-tentacles.html

[25] Lee H, Kwon YC, Kim E. Jellyfish Venom and Toxins: A Review Marine and Freshwater Toxins. , Dordrecht: Springer Netherlands 2015; pp. 1-14. [cited 2017 Nov 15]; Internet http://link.springer.com/10.1007/978-94-007-6650-1_26-1

[26] Brinkman DL, Konstantakopoulos N, McInerney BV, *et al.* Chironex fleckeri (box jellyfish) venom proteins: expansion of a cnidarian toxin family that elicits variable cytolytic and cardiovascular effects. J Biol Chem 2014; 289(8): 4798-812.
[http://dx.doi.org/10.1074/jbc.M113.534149] [PMID: 24403082]

[27] Thaikruea L, Siriariyaporn P. Severe Dermatonecrotic Toxin and Wound Complications Associated With Box Jellyfish Stings 2008-2013. J Wound, Ostomy Cont Nurs [Internet] , 2015 [cited 2017 Nov 15];42(6): 599-604. Available from: http://content.wkhealth.com/linkback/openurl?sid=WKPTLP:landingpage&an=00152192-201511000-00004

[28] Conidae (Cone Shells) [Internet]. [cited 2017 Nov 16]. Available from: http://shells.tricity.wsu.edu/ArcherdShellCollection/Gastropoda/Conidae.html

[29] Haddad V Junior, de Paula Neto JB, Cobo VJ. Venomous mollusks: the risks of human accidents by conus snails (gastropoda: conidae) in Brazil. Rev Soc Bras Med Trop [Internet] SBMT , 2006 Oct; [cited 2017 Nov 16];39(5): 498-500. Available from: Conidae (Cone Shells) [Internet]. [cited 2017 Nov 16]. Available from: http://www.scielo.br/scielo.php?script=sci_arttext&pid=S0037-86822006000500015&lng=en&tlng=en

[30] Cone Shells' (Conidae) Venom Apparatus

[31] Becker S, Terlau H. Toxins from cone snails: properties, applications and biotechnological production. Appl Microbiol Biotechnol [Internet] Springer , 2008 May; [cited 2017 Nov 16];79(1): 1-9. Available from: http://www.ncbi.nlm.nih.gov/pubmed/18340446
[http://dx.doi.org/10.1007/s00253-008-1385-6]

[32] Lee C-C, Tsai W-S, Hsieh HJ, Hwang D-F. Hemolytic activity of venom from crown-of-thorns starfish Acanthaster planci spines. J Venom Anim Toxins Incl Trop Dis [Internet] BioMed Central , 2013 Sep 24; [cited 2017 Nov 16];19(1): 22. Available from: http://www.ncbi.nlm.nih.gov/pubmed/24063308

[33] Shiomi K, Midorikawa S, Ishida M, Nagashima Y, Nagai H. Plancitoxins, lethal factors from the crown-of-thorns starfish Acanthaster planci, are deoxyribonucleases II. Toxicon [Internet] , 2004 Oct; [cited 2017 Nov 16];44(5): 499-506. Available from: http://www.ncbi.nlm.nih.gov/pubmed/15450924
[http://dx.doi.org/10.1016/j.toxicon.2004.06.012]

[34] Lee C-C, Tsai W-S, Hsieh H-J, Hwang D-F. Cytotoxicity of venom from crown-of-thorns starfish (Acanthaster planci) spine. Mol Cell Toxicol [Internet] Springer Netherlands , 2013 Jun 18; [cited 2017 Nov 16];9(2): 177-84. Available from: http://link.springer.com/10.1007/s13273-013-0022-3

[35] US Department of Commerce NO and AA. What is a lionfish? [cited 2017 Nov 16]; Available from: https://oceanservice.noaa.gov/facts/lionfish-facts.html

[36] Beattie M, Nowacek DP, Bogdanoff AK, Akins L, Morris JA. The roar of the lionfishes *Pterois volitans* and *Pterois miles*. J Fish Biol [Internet] Blackwell Publishing Ltd , 2017 Jun 1; [cited 2017 Nov 16];90(6): 2488-95. Available from: http://doi.wiley.com/10.1111/jfb.13321

[37] Red Lionfish | National Geographic [Internet]. [cited 2017 Nov 16]. Available from: https://www.nationalgeographic.com/animals/fish/r/red-lionfish/

[38] "Crocodile Hunter"; Death Extremely Rare, Caught on Film [Internet]. [cited 2017 Nov 16]. Available from: https://news.nationalgeographic.com/news/2006/09/060905-irwin.html

CHAPTER 7

Spiders, their Venoms, and a Bit More

Abstract: Spiders rarely engender neutral feelings in people; many fear them and some find them repugnant. There are a few examples in literature of spiders portrayed in a positive light and they are whimsically described in children's verses and stories. Scientifically, spiders are interesting biologically for their behavior, especially their webs. Spider silk from which webs are made is noted for its high strength, which exceeds (on a weight-comparison basis) steel and man-made fibers, and its elasticity which allows webs to catch flying insects with impacts of a thousand watts of power. Spider venoms have a wide variety of chemical structures and biological activities. Some, including venom of the black widow spider, have neurotoxic components and the complete venom of this spider is similar in toxicity to that of rattlesnake venom. The Brazilian wandering spider and the Australian funnel web spider vie for the title of most venomous spider. The brown recluse is feared over large regions of the central United States because of the large necrotic wound that can result from their bite and their reclusive nature coupled with their tendency to occupy human residences. The tarantula is widespread around the world and has unusual ability to shoot poisonous hairs from its body in addition to a venomous bite which, fortunately, is not usually medically serious for humans. Venoms from many spider species are useful for scientific studies because some interfere with the mechanisms used for communication between and within cells for various physiologically essential functions. Spider venoms are being investigated as tools for studying nerve cell functions including impulse transmission. They also are being explored as pain killers and used as tools in the search for causation and cures for several devastating neurological conditions.

"When I get sick of what men do, I have only to walk a few steps in another direction to see what spiders do. Or what the weather does. This sustains me very well indeed"[1]

Keywords: Agatoxins, Ampullate, *Atrax Robustus*, Black Widow, Brazilian Wandering Spider, Brown Recluse, Brown widow, Calcium Channel, Cysteine Knot Toxin, Delta-altracotoxins, Funnel-web Spider, Grammatoxin, Neurotransmitter, Potassium Channels, Spider Silk, Spider Web, Tarantella, Tarantism, Tarantula, Vanillotoxins.

INTRODUCTION

When first I think of spiders, I like to think of their beautiful and intricate webs, sometimes glistening in the sun and sometimes laden with drops of dew. The webs are strong but succumb to our annoyed flailing that destroys their night's

Olen R. Brown

work in a single gesture. Most spiders are our friends, but because of our fear of the few, we lump them all into a pile of unknown fears, much as we sometimes do amiss with people we don't know.

I choose to begin with the web and not the spider. Let us first consider an example of their industrious work and do so without focusing on our fear of their poisons which are influenced, rightly or wrongly, by our earliest experiences from childhood.

THE SPIDER'S WEB

Spider webs are unique in nature. It is a wondrous work that combines chemistry, art, and industry. Perhaps the spider, in addition to the ant, deserves credit for its work ethic. Spiders are sometimes even classified by the fact that they spin webs which are used for catching insects as opposed to hunting spiders that chase their prey. However, some hunting spiders spin simple webs along the ground to aid them when chasing insects. Spider webs are made of silk which the spider spins using silk glands which are complex organs with seven pairs of spinnerets, each different in function. Spiders must make several kinds of silk: silk for attachment; dragline silk; orb web silk that is stretchy; sticky, catching silk; swathing silk for wrapping prey; tangling silk for catching prey; protective egg sac silk; and silk for "ballooning" where a strand of silk is used by baby spiders to aid in their dispersal [1].

Spider silk is biosynthesized as a liquid but emerges from the spinnerets as solid silk fibers. Spider silk is only approximately 0.000039 inch to 0.000157 inch in diameter but stronger than steel of the same dimension [1]. It is primarily a protein. The silk of the Nephila spider is the strongest natural fiber known and is used to make tote bags and fish nets. In certain unusual species, spider silk is used to create a bag for capturing air in a bubble that the spider uses to survive underwater. Web-spinning spiders use only the tips of their middle claw and bristles on their legs when spinning their webs and their bodies do not contact the sticky web. The most common web design is called the 'orb web' and is wheel-like in appearance with an outer frame-line, radial lines and spiral lines [1]. The spiral lines are created last and the spider begins at the center and moves outward using sticky silk which it avoids in this process. These webs can be quite large and their creation follows a specific intricate pattern. There are many other types of webs (Fig. **1**), including tangled web, sheet webs, gum-footed webs, horizontal line webs, and many others that are created by different types of spiders [1]. Indeed, more than 130 different shapes of spider webs have been described [2].

A review of spider silk [2] elaborates details of its structure and function. Spider silk consists almost entirely of proteins and "... often exceeds man-made materials in … properties… Silk fibers have tensile strengths comparable to steel and some silks are nearly as elastic as rubber on a weight to weight basis. In combining these two properties, silks reveal a toughness that is two to three times that of synthetic fibers like Nylon or Kevlar. Spider silk is also antimicrobial, hypoallergenic and completely biodegradable" [2]. From pioneer days there are accounts of using spider webs to cover a wound and stop bleeding.

More than 34,000 different species of spiders are known and approximately 50% use webs to catch prey [2]. The previously described types of silk used in various parts of web-making are very different (Fig. **1**). The frame and radii of orb webs are made of strong, rigid silk made of two types of protein derived from the major (ampullate) silk-producing glands and it is called MA silk. MA silk is also used to make lifelines as escape routes to avoid predators. The spiral threads are spun from special glands and are made of only one kind of protein. This silk is very elastic and readily dissipates the impact energy of flying prey insects. This review [2] states: "… a typical honey bee with a body weight of 120 mg and a maximum flight velocity of about 3.1 meters per second crashes into a spider's web with a kinetic energy of approximately 0.55 mJ (a reference is cited)". One mJ is 1,000 Joules and is equivalent to 1,000 Watts of power dissipated in one second. Prey, larger than the spider, can be captured.

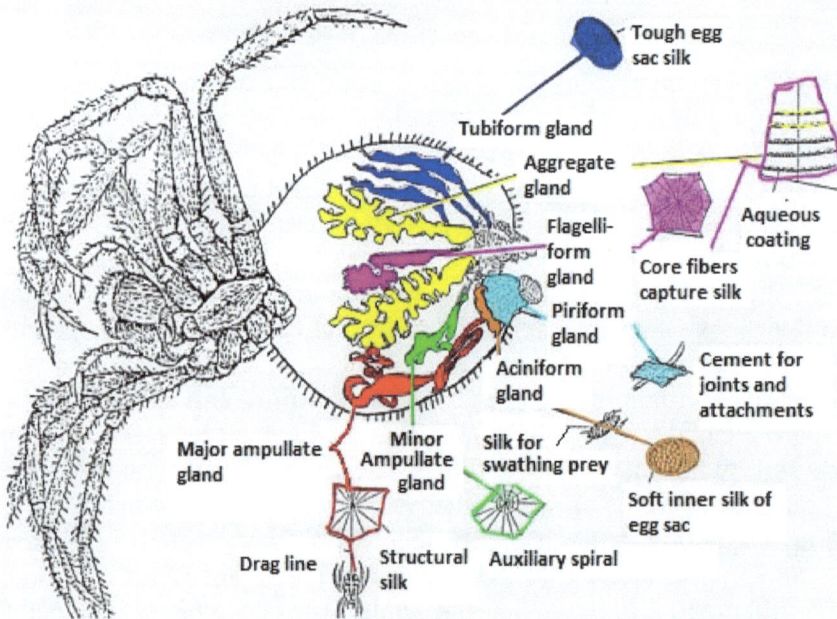

Fig. (1). Spider silk is spun from proteins into various silks for specific purposes.

Orb weavers also make two other types of silk that are used to make their webs. Spiders require silk for web scaffolding connections, attachments to link the web frame to trees, and material for egg case and for other purposes (Fig. **2**). Significantly, some of this silk is purposefully not sticky, and to attach prey to the web additional glue is applied to special threads called capture threads [2]. The chemical and physical structure of spider silk has been examined in detail and described in a review [2] and in a paper published in 2010 [3]. The proteins of spider silk contain large amounts of non-polar and hydrophobic amino acids like glycine and alanine but very little tryptophan. Silk also contains highly-repeated sequences of the same amino acids and these can comprise more than 90% of the entire spider silk protein [2].

Each of these repeating structures contribute to the known outstanding mechanical properties of the silk threads has a distinct function. After secretion from glands, silk proteins are in solution and complex interactions at the chemical level determine the ultimate structure assumed by the silk threads. Assembly of the protein molecules from the highly-concentrated solution occurs automatically in a wondrous process dictated by the chemistry of molecular bonding and subsequent folding into particular three-dimensional protein structures [2]. This original source should be consulted for those interested in protein chemistry.

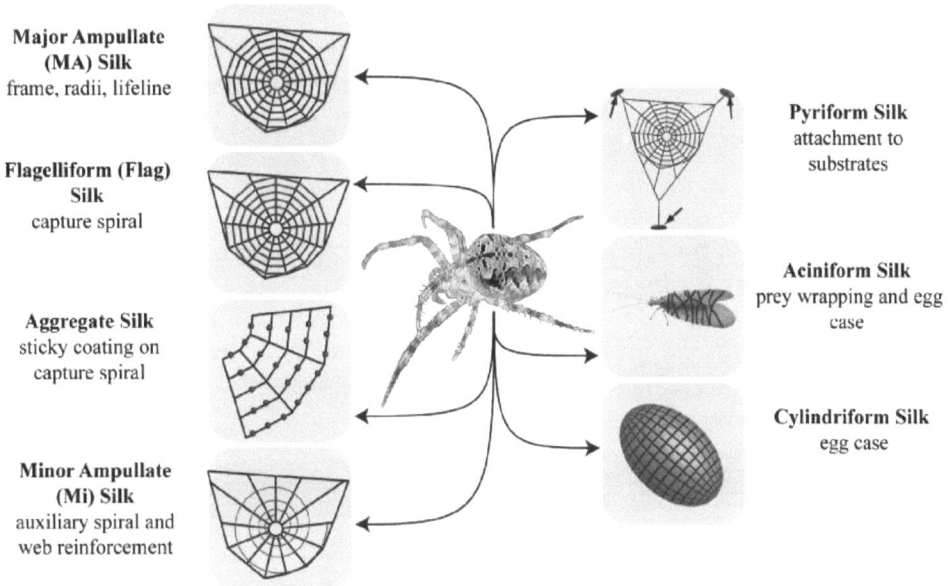

Fig. (2). Spider silk is made for specific purposes.

The mechanical properties of spider silk, previously alluded to, are exceptional and further comparisons are warranted. Spider silk has been tested and shown to absorb three-times more energy on a weight-to-weight basis than Kevlar, one of the sturdiest man-made materials [2]. Spider silk has been compared to silk taken from the cocoons of pupae of the silkworm *Bombyx mori* and "… differences between insect and spider silks are evident on all levels, from the molecules involved, to the structural arrangements of the proteins, to the mechanical properties of the thread… depending on spinning conditions, silkworm silk is either strong or elastic, whereas spider silk combines both properties" [2].

SPIDERS IN LITERATURE

The word *Arachne*, a classification term for spiders, has its origin in Greek mythology. The word spider comes from words that refer to spinning and it is an ancient and powerful symbol that, even today, evokes strong emotions than range from disgust and fear, to curiosity, and to humor (in children's rhymes). Spider has even become a stock market symbol. Spider, spelled differently (Spyder, or sometimes spelled the same), is a name for an exotic automobile.

The Lakota tribe of American Indians have stories perpetuated by oral tradition that include a tale of a spider who is a trickster chronicled in their legend about the 'Wasna' (pemmican) man and the 'Unktomi' (spider) [4].

Ovid, Publius Ovidius Naso, [43 B. C.- A.D. 17/18], perhaps best known for his 15-book mythological compendium *Metamorphoses,* wrote in the 6th book about rivalry between gods and mortals which begins with Arachne a tale about a spider [5].

Charlotte's Web by E. B White portrays the spider in a positive light; indeed, as the heroine. In this classic, the main character is a young pig who is about to be slaughtered and eaten for Christmas dinner. The story is about his friendship with a barn spider named Charlotte [6].

A most notable children's verse, by an anonymous author lost to history, implies that the spider is a universally feared, but quite misunderstood, creature (Fig. **3**). Who does not recall having been read this verse as a child?

> "Little Miss Muffet who sat on a tuffet, eating her curds and whey; Along came a spider, who sat down beside her, and frightened Miss Muffet away"[2].

The Spider and the Fly is a classic poem by Mary Howitt. It is a story of seduction and betrayal that was the inspiration for a 1965 Rolling Stones song and a 1923 Cartoon by Aesop's Fables Studios [7]. Here is the poem:

"Will you walk into my parlour said the Spider to the Fly,

'Tis the prettiest little parlour that ever you did spy;

The way into my parlour is up a winding stair,

Oh no, no, said the little Fly, to ask me is in vain,

For who goes up your winding stair - can ne'er come down again.

And I've a many curious things to shew when you are there."

Fig. (3). Little Miss Muffett sat on a tuffet.

Walt Whitman describes a 'ballooning' spider in a poem he wrote in 1868: "A Noiseless, Patient Spider" [8]. This use of spider web as a means of dispersing baby spiders was previously mentioned in this chapter. Whitman was a naturalist and apparently was impressed by this characteristic of some spiders.

For Robert the Bruce of Scotland, the spider was a symbol of inspiration according to legend; however, this may be only apocryphal. In the legend, he encounters a spider during the period when he has a series of military failures fighting against the English. In one version, he shelters in a cave and witnesses a spider continually failing in attempts to climb a thread to reach its web. With great perseverance at last the spider succeeds. Robert takes this as a sign and by great effort independence eventually was won by Scotland. This and a wealth of other legends are found in 'Spiderlegends at Spiderzrule' [9].

Athena and Arachne

Arachne, in Greek mythology, was a weaver (and a mortal) who challenged Athena (a god), and was subsequently turned into a spider. This myth has at least three versions [10, 11]. In one version, Arachne is a beautiful shepherd's daughter who is particularly skilled at weaving. Boasting, pridefully, of her skill infuriated Athena who arranged a contest of skills. Athena weaved four scenes in which the gods were depicted punishing those humans who felt they were equal to gods; the sin of hubris. Arachne wove scenes in which gods abused humans. Arachne's art was clearly better than Athena's. Athena cast a powerful potion, devised by Hecate, onto Arachne and she was transformed into a spider and was condemned to weave eternally. This provides the connection with the origin of Arachne as a classification term for spiders.

Another version begins similarly; however, during the challenge, Athena realizes the skillfulness of Arachne. Frustrated, she wants to teach her a lesson by shaming her. Arachne's shame was so great that she hung herself. Athena brings her back to life as a spider and she is allowed (or destined) to weave eternally. In a third version, Zeus was the judge in the contest between Arachne and Athena. The loser would not be allowed to touch the loom again. Athena won the contest in this version. Arachne was devastated and by oath she could never weave again. Out of pity, Athena transforms her into a spider so she could continue weaving without breaking the oath.

Other Spider Tales

The tarantula has been portrayed in many legends and myths. According to Celtic tradition, this spider was a beneficial being or a sacred creature in stories that were written or only passed down orally [12]. This site does recount one Celtic tale said to be found in *Folk Tales of Brittany* written by Elsie Masson in 1929 which I quote from as follows because there is no other way adequately to tell the story.

> "Two brothers are traveling through a forested countryside... [they meet] a beggarly old woman. The first brother ignores the woman, while the second gives hers all of his coins... the old lady gives the generous brother a walnut, which she claims contains a wasp with a diamond stinger. The older brother [is] annoyed at his younger brother's generosity, but opts to continue the journey... the brothers come across a boy who's shivering in the wind. The younger brother gives the boy his cloak... much to the dismay of his older brother. The child [boy]... gives him a dragonfly... The brothers continue on their way... they come upon an old man who says that he cannot walk. Filled with compassion, the younger brother gives the old

man his horse. In return, the old man gives the generous brother a hollow acorn… which he claims contains a spider… Angry at his younger brother's foolishness, the older brother leaves him cold, penniless, and without a mount… a giant eagle snatches him [the older brother] from the saddle and carries him into the clouds. The younger brother, horrified… witnesses the act and sees that two eagles have committed it. The younger brother… wonders aloud… how he will be able to rescue his older brother… he hears three tiny voices… [that] beg to help him… the younger brother unleashes the three deadly insects [spider, dragonfly, wasp]… the spider springs forth and immediately weaves a ladder towards the heavens from upon the dragonfly's back. The group then ascends into the sky, and into the lair of the giant responsible for the kidnapping… A battle of epic proportions ensues. The wasp stings out the eyes of both eagles and the giant. The spider then attacks, and wraps the giant within its steel-like web. Suddenly, the eagles switch sides… blindly pecking the giant to death through the spider's binding web. The eagles both die on the spot… In the after math of carnage, the dragonfly and wasp transform into horses and attach themselves to [a] reed cage which has now become a coach. The spider… has become the carriage's groom. The brothers get in and the coach carries them through the sky [and] to their horses, where the younger brother … finds a much fuller coin purse and a diamond-studded cloak… The brothers fall to their knees and thank the divine beings for all of their aid and for saving their lives."

This account also describes various other appearances of spiders in writings including in the book *Notes on the Folklore of the Northeast of Scotland* written in 1881 by Waltor Gregor. Some details from this book as related in an account [12] are difficult to portray other than as quotations: "Spiders were regarded with the feeling of kindness, and one was usually very loath to kill them. Their webs, very often called 'moose webs', were a great specific to stop bleeding… A small spider makes its nest – a white downy substance – on the stalks of standing corn. According to the height of the nest from the ground was to be the depth of snow during winter."

For more Celtic tales involving spiders, read the website *Folk Tales of Brittany* [12].

BLACK WIDOW

Spiders (the order *Araneae*) are reported to be the largest group of venomous organisms [13]. The female black widow spider (*Latrodectus species)* has a striking appearance. She has a shiny black color with a red, hourglass-shaped design on the underside of her abdomen (Fig. **4**). This spider is notorious for cannibalizing its mate which is about half her size. Sources make the claim that it is the most venomous spider in North America [14], or that it is one of the most venomous spiders in the world [15]. Its bite is rarely fatal, however, and an antivenin is available [16]. Sometimes, the hourglass is an orange-yellow color. There are 31 species including *Latrodectus hersperus,* (the western black widow), *Latrodectus mactans* (southern black widow), and *Latrodectus variolus (northern black widow)* that are referred to as 'black widows'. There are also 'red widows', and "brown widows" (Fig. **5**) that are usually described as less toxic [14]. Females are said to sometimes cannibalize the male after mating and this is the source of a part of the spider's common name. Females have papery-like egg sacs with 200 to 900 eggs that hatch in about one month. The babies are cannibalistic and few of the hatchlings survive the 3-month period needed to mature [14].

Fig. (4). Black Widow spider, showing typical hourglass-shaped red spot on abdomen, with egg sac in a delicate spider's web.

Sometimes black widows are referred to as comb-footed spiders because they have a series of stiff, short bristles, that resemble the teeth of a comb, on the last segment of their fourth pair of legs [14]. The female is approximately 1.5 inches in length and the male is half this size. This spider is found in temperate regions around the world. In the U.S. it is primarily in the South and West [14]. From personal knowledge I know that it is found in Oklahoma and Missouri. They can be found in barns, garages, basements, outdoor toilets, hollow stumps, rodent holes, trash, brush, and dense vegetation [14, 17]. When I was a small boy in Oklahoma, I was cautioned about this spider and recall an occasion when I was in the barn of a neighbor and he began shaking his head and said I was to be wary of black widows because they liked to nest in his barn which was full of hay. Whether it was real or imagined, I recall, even today, that I felt something on my neck and raked at it and saw in my palm a small black spider which I immediately flung away. I never learned if it was really a black widow, but if it had been, it was either a male or immature female and I was not in real danger.

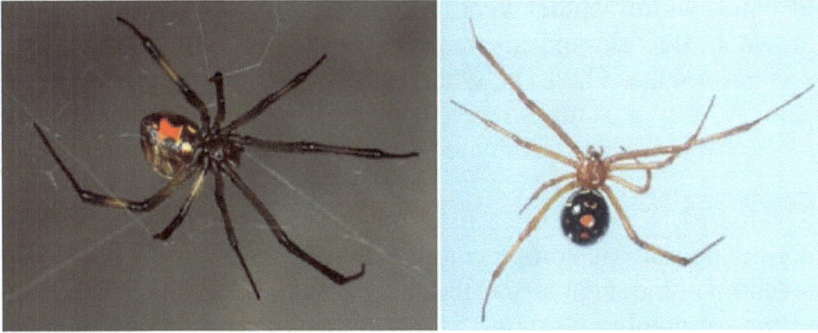

Fig. (5). Brown widow spider (left) and Red widow spider (Right).

Black Widow Venom, α-latrotoxin plus a Complex of Enzymes

The black widow spider produces what may be the most complex venom known. Evolutionists are not able to explain why this small insect produces a complex toxin that can kill an animal as large as a human. It seems to make no sense that this spider would produce venom toxic enough to kill even a small animal such as a mouse. Perhaps the answer is that the enzymes in the venom help to dissolve the tissues of the spider's prey which then can be sucked into the spider in a liquid form with pre-digestion already initiated. In order to understand how this venom works, it will be necessary to describe some biochemistry that involves details and perhaps unfamiliar terms. As these details are presented, keep in mind that in simplest terms the venom kills insects and initiates the liquefaction of their tissues preparatory for ingestion, but also for unknown reasons is toxic enough to kill a human. Additionally, some of the toxins are found outside the venom glands in baby spiders and in egg sacks for reasons that are speculative.

A recent review of widow spider venoms [18] states that these toxins "… have received much attention due to the frequently reported human and animal injuries caused by them". This review also reports that the knowledge of the mechanism of action of these toxins is still somewhat unclear. Also, unlike most venomous creatures, the venom is not only in the venom glands but also in other body parts of 'newborn' spiders, and in the spider's eggs. A systematic investigation of this toxin has identified four protein-like toxic components. The venom secreted by the spider's gland is a complex mixture with several physiologic actions on other animals including humans. The toxin is composed of biologically active proteins and peptides. Peptides, like proteins, are made of amino acids but they are much smaller molecules. These toxins have many effects on the prey of spiders including the abilities to paralyze, immobilize, kill, and liquefy.

One toxin component at a very low concentration was found specifically to increase Ca^{2+} influx into certain cells [18]. This review also reports that a study found that black widow spider venom contained large proteins with molecular weights above 10,000 Da and most were in the 100,000 Da range. This venom was rich in neurotoxins. The $LD_{50,}$ when injected into mice, was 0.16 mg of toxin per kg of mouse. A typical mouse weighs approximately 20 grams; therefore, only 3.2 µg would kill a mouse. Calculated for the human (simplistically, on a body weight basis) this equates to 11.2 mg as the lethal dose for the average human [18]. This value (0.16 mg/kg) is a little more poisonous that the value typically reported for the neurotoxic component of rattlesnake venom, 0.2 to 0.7 mg per kg [19]. This venom component also effectively blocked neuromuscular transmission in isolated mouse diaphragm while a low molecular weight component (smaller than 10 kDa) was ineffective which indicated that the toxicity of the venom for mammals was primarily associated with a large protein component. This review [18] also states that the venom contains "multiple kinds of hydrolases, including proteinases, hyaluronidase, and alkaline and acid phosphatases. All these data demonstrate that the venoms are rich in ion-channel modulators and metabolic enzymes, particularly the proteolytic enzymes that can enhance the action of the toxins by breaking down the intercellular reinforcements and basement membrane molecules."

An article in 2010 [20] summarizes information known about α-latrotoxin at that time and describes its target (Fig. **6**) as "release sites of neurotransmitters and mediators at presynaptic terminals in neurons and secretory cells where it induces massive and uncontrolled exocytosis of the carrier vesicles. Molecular mechanism of α-latrotoxin is complex and not fully understood." The general sequence of events as described in the 2015 review [18] were known.

This review [18] further states that recent advances in protein separation techniques and means of mass analysis has permitted assessment of "… virtually all venom components". This review cites a paper that amazingly reported that a total of 122 distinctly different components were identified from venom obtained by electrostimulation and that 75 of these appeared to have distinct toxic functions. "Besides the previously reported widow spider venom proteins, including latrotoxins, a variety of hydrolases and other proteins with special activity were found in the venom, such as phosphatases, nuclease, fucolectin, venom allergen antigen 5-like protein, and trypsin inhibitor." This review [18] also cites research stating: "Sixty-one proteins were identified… from a *Latrodectus Hesperus* database that matched peptides collected from *Latrodectus Hesperus* venom, including 21 latrotoxins, one ICK (inhibitor cysteine knot) toxin, and six CRISP (cysteine-rich secretory protein) family toxin proteins. Several types of enzymes were identified in the venom, including hyaluronidases, chitinase, serine proteases, and metalloproteases."

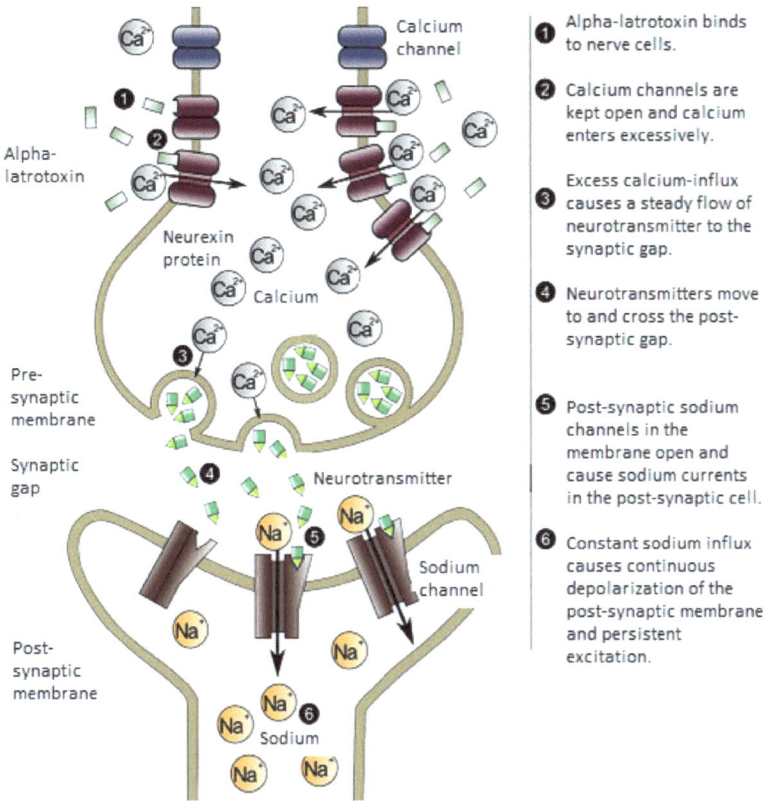

Fig. (6). Site and mechanism of action of α-latrotoxin on human nerve cell impulse transmission at the myoneural junction.

Recently developed DNA sequencing procedures and associated techniques have been applied to the complex problem of identifying the many toxic components produced by black widow spiders [18]. They report that 146 proteins were demonstrated, with high confidence, to be toxin-like proteins forming 12 families which are described in a table in this review which can be consulted for details.

The high molecular weight components of black widow spider toxins were also described earlier (in 1998) by E. V. Grishin [21]. The ionophoric and secretogenic actions of the toxins were described from highly purified α-latrotoxin, and they also reported the presence of a low molecular weight protein that was structurally related to crustacean 'hyperglycemic hormones'. Calcium-dependent and calcium-independent binding sites for the toxin were found in presynaptic membrane preparations.

At least seven different latrotoxins have been isolated and all are large, acidic proteins with masses from 110,000 to 140,000 Da [18]. Most of these proteins act against insects and are termed latroinsectotoxins. The only known venom component of this spider that is targeted specifically at vertebrates is α-latrotoxin according to Yan and Wang [18]. One toxin is active only against crustaceans. All of these latrotoxins cause massive release of neurotransmitters from the nerve terminals of the retrospective species after binding to specific neuronal receptors. However, this review also reported that two low-molecular-weight proteins (called latrodectins) are inactive against mammals and insects. It was speculated that these proteins augment the neurotoxicity of latrotoxins by increasing their affinity for the membrane target and by reducing toxicity specifically aimed at vertebrates with the effect that α-latrotoxin becomes effective for insects. This review, in 2015 [18], states: "The data available indicate that *Latrodectus* spp. venoms contain a cocktail of toxins and other biologically active substances; however, many of them wait to be isolated and characterized."

The most-studied latrotoxin is α-latrotoxin which has a molecular mass of about 130,000 Da [18], and its lethal action in vertebrates is to cause massive release of neurotransmitter [22] (Fig. **7**). It has been learned that α-latrotoxin creates Ca^{2+} -permeable channels in lipid bilayers, and that this involves toxin assembling as tetrameric complexes around a central membrane channel [22]. The increased Ca^{2+} influx through the channels in the presynaptic membrane is responsible for much of the effects of this toxin. Three proteins in the plasma membrane, including neurexin (Fig. **7**) and an enzyme have been shown to function at the receptor site for α-latrotoxin [18]. The biochemical details of this binding have been investigated and reported [23]. After binding to a receptor, α-latrotoxin exerts its effect by two major mechanisms. One is Ca^{2+} -dependent action to insert the toxin into the plasma membrane with pore formation and the other is Ca^{2+} -independent based on receptor-mediated signaling. There are three specific receptors for α-latrotoxin; however, neurexin binds to the toxin only in the presence of extracellular Ca^{2+}. The authors of the review [18] state: "In the past decades, latrotoxins, particularly α-latrotoxin from the venom of the black widow spider have been extensively employed as tools to investigate the molecular mechanisms involved in the regulation of neurotransmitter release." The toxin has also been evaluated for its ability to antagonize botulinum poisoning. The authors of this review [18] further state: "The botulinum neurotoxins… are the most poisonous substances known and exhibit zinc-dependent proteolytic activity against members of the core synaptic membrane fusion complex, preventing neurotransmitter release and resulting in neuromuscular paralysis… α-latrotoxin attenuated the severity or duration of [botulism neurotoxin-] induced paralysis in neurons."

The black widow toxins that are present outside the venom glands have been the target of research interest. They are called non-venom toxins [18]. These toxins are found in the eggs of the spider and in one species they were found to be rich in high-molecular mass proteins and in peptides below 5,000 Da that showed toxicity to mammals and insects, and a total of 157 proteins were identified from egg extracts. These toxins showed various impairing activities against important cell functions. The protein composition of the toxic egg extract appears to be even more complex than that of venom and "… there are only a few similarities between the protein compositions of the two materials. No known typical black widow spider venom proteins were found in the egg extract, suggesting that the eggs have their own distinct toxic mechanism" [18]. Recently-hatched black widow spiders show obvious toxicity to animals [18]. Aqueous extract of "newborn" spiders, when injected into mice and cockroaches caused poisoning symptoms and deaths. The LD_{50} was 5.30 mg per Kg and 16.74 µg per g, respectively, in the mouse and cockroach.

There has been speculation about implications of toxins present outside the venom glands in black widow spiders. Could these toxins function to provide protection against aggressive arthropods? A reference for this is provided in the review which we have extensively cited [18]. As speculation it is also said that: "… the antibacterial components may play important roles in protecting the eggs and newborn spiderlings from some pathogenic microorganisms."

SYDNEY FUNNEL-WEB SPIDER

A paper published in 2012 [24] states: "Australian funnel-web spiders are generally considered the most dangerous spiders in the world, with envenomations from the Sydney funnel-web spider *atrax robustus* resulting in at least 14 human fatalities prior to the introduction of an effective anti-venom in 1980." There are approximately 40 species of funnel-web spiders placed in the genera *Hydronyche* and *Atrax* [25]. They are medium to large spiders (approximately ½ to 2 inches in body length) (Fig. **7**).

Fig. (7). Australian Funnel Web Spider.

"Body color can vary from black to brown but the hard carapace covering the front part of the body is always sparsely haired and glossy… not all species are known to be dangerous, but several are renowned for their highly toxic and fast-acting venom. The male of *Atrax robustus,* the Sydney Funnel-web Spider, is probably responsible for all recorded deaths by this species (13) and many medically-serious bites. This remarkable spider has become a part of Sydney's folklore and, although no deaths have been recorded since the introduction of an antivenom in 1981, it remains an icon of fear and fascination for Sydneysiders." [25]. Take note that a previously cited reference [24] has a slightly different date for introduction of antivenom, but the same excellent results.

"Funnel-web spiders live in the moist forest regions of the east coast and highlands of Australia from Tasmania to north Queensland… Funnel-webs of the genus *Atrax* have a much smaller distribution… The Sydney Funnel-web Spider, *Atrax robustus,* is found from Newcastle to Nowra and west as far as Lithgow in New South Wales." [25]. Funnel-webs are described to create burrows in the ground and even in rotting logs and insect bore-holes in rough-barked trees. They rarely are found in open areas. "The most characteristic sign of a Funnel-web's burrow is the irregular silk trip-lines that radiate out from the burrow entrance of most species. These trip-lines alert the spider to possible prey, mates or danger… The spider (hunting mostly at night) sits just inside the entrance with its front legs on the trip-line. When a beetle, cockroach, or small skink, typical items of funnel web food, walks across the lines, the spider senses the vibrations and races out to grab its meal. The prey is quickly subdued by injection of venom from the spider's large fangs. Funnel-web spiders also forge on the surface in the vicinity of the burrow" [25].

"Only male spiders have been responsible for all recorded funnel-web envenomation deaths" [25]. This fact also is found in the previously sited report [13]. This results from a variety of factors including venom chemistry and spider behavior. The male spider venom contains "… a unique component called robustoxin (d-Atracotoxin-Ar1) that severely and similarly affects the nervous systems of humans and monkeys, but not of other mammals" [25]. This venom subsequently will be further described.

Spiders are said to be "… the most megadiverse animal group on the planet… with 3,859 genera comprising 42,752 species having been described … only six genera of spiders are capable of inflicting lethal envenomations in humans." The Australian funnel-web spiders (*Araneae*: *Mygalopmorphae*: *Hexathelidae*: *Atracinae*) are a group of approximately 40 species that comprise three of these lethal genera: *Atax, Hadronyche*, and *Illawarra.* Representatives from the genus *Atax* caused at least 14 human deaths between 1927 and the introduction of antivenom in 1980 [24]." Most recommendations are that all suspected bites of humans by any funnel-web spider should be considered as potentially dangerous. "Besides *Atrax robustus* several other species have been sporadically involved in life threatening envenomations." They include the Blue mountain Funnel-web spider (*Hadronyche versuta*) and the Southern and Northern Tree Funnel-web Spiders (*H. cerbera* and *H. formidabilis*) [25].

An interesting bit of side information: funnel-webs can't jump, but they can run quickly and they can rear up and make sudden lunges. When they fall into backyard swimming pools they can stay alive for hours; however, they cannot swim. They can trap a bubble of air in hairs around the abdomen which helps them float and can be used as an air supply when submerged. They have been known to survive submerged for 24 to 30 hours. It is wise not to assume that a spider at the bottom of a pool is dead [25]. A near fatal envenomation of an infant by a funnel-web was reported [26]. Another study reported results of 16 funnel-web bites [27]. Ten cases had only minor, local effects; four had moderate envenomation (non-specific systemic or local neurotoxicity), and two had severe envenomation that required antivenom [27]. Acute myocardial injury was reported to be caused by a bite by the Sydney funnel-web spider (*Atrax robustus*) [28]. The same report recounts that a 67-year-old woman was bitten by a Sydney funnel-web spider. She had signs consistent with acute myocardial injury, was treated with funnel-web spider antivenom, given intensive care treatment with initial respiratory support for acute pulmonary edema, and she survived. There were indications that the patient may have had minimal atherosclerotic disease [28].

Clinical features and management of five cases of Australian funnel-web spiders has been published [29]. The cases were: "… the first life-threatening envenomation by *Hadronyce* species 14 (the Port Macquarie funnel web). Two severe envenomations by *Hadronyche cerberea* (the Southern Tree funnel web) and one each by *Hadronyche formidabilis* (the Northern Tree Funnel web) and *Hadronyche infensa* (the Darling Downs funnel web) are also described." Several cases from the literature were also described. "It is concluded that bites from at least six *Hadronyche* species have produced a life-threatening envenomation syndrome clinically indistinguishable from that of *Atrax bombustus*. *Atrax bombustus* derived antivenom is effective although antivenom requirements may be greater than for *Atrax* envenomation" [28].

Funnel-Web Spider Venoms

Three main syndromes result from spider bites: latrodectism (widow spiders), loxoscelism (spiders of the *Sicariidae* family), and funnel-web spider envenomation [30]. Funnel-web spider venom has neurotoxins that stimulate release of neurotransmitter to cause sensory disturbances and muscle paralysis. Proper management of the envenomed patient, including prompt transport to the hospital, correction of the hemostatic disorder, ventilator support, and administration of antivenom significantly reduce the risk of neurological complications which, in turn, reduce the mortality and improve the functional outcome of survivors" [30].

"Delta-altracotoxins (delta-ACTX) (Fig. **8**), isolated from the venom of Australian funnel-web spiders, are responsible for the potentially lethal envenomation syndromes seen following funnel-web spider bites [31]. These venoms are polypeptides made-up of 42 amino acids with four disulfide bridges and what is described as the molecular structure 'inhibitor cystine-knot' that has structural similarity, but not DNA sequence homology, with other spider and marine snail venoms [31]. The altracotoxins cause spontaneous and repetitive activation of nerve impulse transmission and cause prolonged nerve action potentials. This results in neurotransmitter release from muscle and autonomic nerve endings. As a result, there is slowing of voltage-gated sodium channel inactivation and a hyperpolarizing shift of the voltage-dependence of activation [31, 32]. This action is similar, but not identical, to the action of scorpion alpha-toxin and sea anemone toxins which were previously described in this book. The inhibitory effect on respiration of funnel-web spider venom is due to the potent neurotoxins [33].

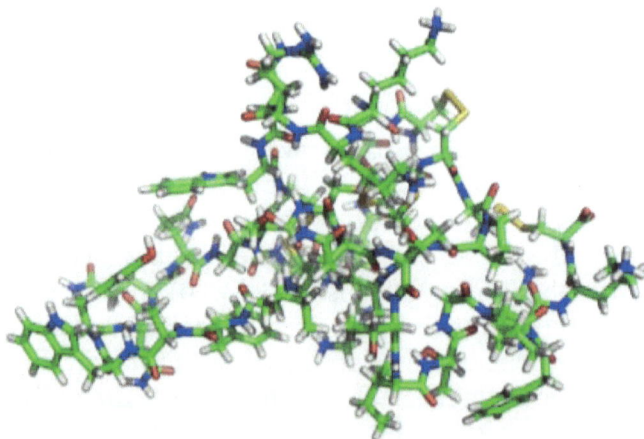

Fig. (8). Model of Δ-atracotoin, the venom from funnel-web spider.

Agatoxins are present also in some American funnel-web spiders. These toxins target three classes of ion channels: transmitter-activated cation channels, voltage-activated sodium channels, and voltage-activated calcium channels [34]. These toxins produce rapid but reversible paralysis in insect prey. Other types of these toxins (alpha-agatoxins and omega-agatoxins) modify both insect and vertebrate ion channels; micro-agatoxins are selective for insect ion channels [34, 35].

A paper in 1998 [36] characterized robustoxin, the lethal neurotoxin from the Sydney funnel-web spider. This rather early work, indicated that "robustoxin inhibits conversion of the open state to the inactivated state of tetrodotoxin-sensitive sodium channels, thus allowing a fraction of the sodium current to remain at membrane potentials at which inactivation is normally complete... robustoxin should henceforth be known as delta-altracotoxin-Ar1 to reflect this main action on sodium channel inactivation. These present results further support the hypothesis that funnel-web spider toxins interact with neurotoxin receptor site 3 to slow channel inactivation in a manner similar to that of alpha-scorpion and sea anemone toxins." This work is historically important because of the data presented that allowed the two statements quoted last (immediately above).

There is a story in the Bible[3] of events occurring in a cave when David is being pursued by King Saul. A similar account, with different details involving a spider's web, is given in a satirical work "*Alphabet of Ben Sira*" [37]. David is saved when Saul sees a spider's web over the entrance to a cave that dissuades Saul and his men from entering to search because the unbroken web indicates that no man could have entered the cave. David is actually hiding in the cave and the spider spins the web to hide his entry. The *Alphabet of Ben Sira* is described as "... one of the earliest, most complicated, and most sophisticated Hebrew stories written in the Middle Ages. Four versions of the work [are known]... All... share a special, satirical, and even heretical, character..." Ben Sira is a person and "Alphabet" refers to 22 alphabetically arranged epigrams attributed to Ben Sira. The fourth part of the account includes the story and is communicated by Ben Sira's son and his grandson. It is said that "Maimonides and other authorities attacked the work vigorously, but it is generally accepted as part of the midrashic tradition" [37].

BRAZILIAN WANDERING SPIDER

The Brazilian wandering spider (Fig. **9**), also known as the armed spider and the banana spider, has the genus name *Phoneutria* which means "murderess" in Greek [38]. It is large, and its body that can be up to 2 inches across. Most are brown; however, its color may vary. The spider is hairy and there may be black spots on the underbelly. Various sources all refer to this spider in similar terms: "... one of the most venomous spiders on earth". Bites of this spider are said sometimes to be deadly to humans, especially children but death is unlikely if antivenom is given. This source [38], citing a scientist, states that 0.006 mg is the lethal dose for a mouse. Calculations (based simplistically on body weight comparisons) show that this venom has a toxicity of approximately 0.3 mg per kilogram and 21 mg would likely kill an adult human. Compared to other toxins, it has a potency similar to that of the neurotoxin in rattlesnake venom (*Crotalus* spp.) which has a LD_{50} of 0.2 to 0.7 (mg per kg, for the mouse) [39]. The Guinness Book of World Records has named it the world's most venomous spider in 2010 and in other years [38, 40].

Brazilian wandering spiders are said not to build webs to catch prey but rather they ambush or directly attack their food which includes other spiders, insects, mice, and small amphibians and reptiles [38]. This site also states that: "When threatened, they will raise their first two pairs of legs" which may expose scarlet hairs surrounding their fangs in some species.

All eight known species of this spider are found in Brazil; some species are also found in Latin America from Costa Rica to Argentina [38]. This site and other sources commonly report stories that Brazilian wandering spiders have been found in exported bananas – in some of these accounts other spiders (such as *Cupiennius* species, which is said to resemble the wandering spider) may have been misidentified as the wandering spider. Other sources, however, apparently verify the stated information about this spider [41, 42].

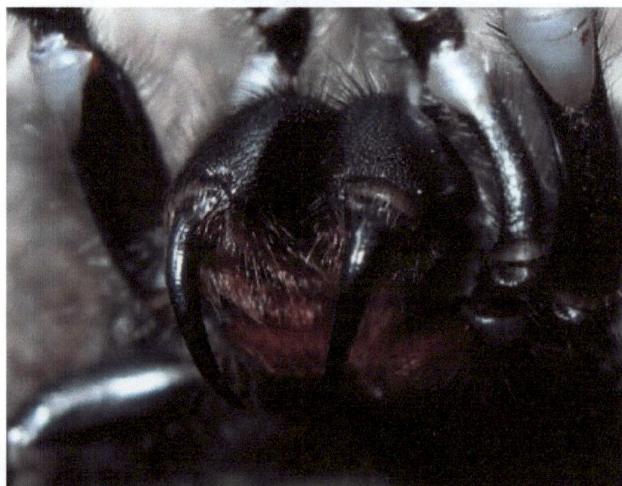

Fig. (9). Brazilian wandering spider (Brazilian banana spider) showing the large fangs which are described to be capable of piercing shoe leather (which may be an exaggeration).

Effects of Venom

A curious effect, with possible medical applications, is the fact that male victims of bites sometimes experience a prolonged and painful erection medically termed priapism [40]. This effect probably results from a known action of the toxin to elevate nitric oxide, which in turn increases blood flow to organs. The venom is a complex 'cocktail' primarily composed of proteins and polypeptides. A significant site and mechanism of toxicity is exerted on nerve and muscle cells at the neuromuscular junction (as has been previously reported for other bio-neurotoxins). Signs and symptoms in envenomated victims include: burning pain at the envenomation site, sweating and goose bumps, and within 30 minutes high or low blood pressure, fast or slow heart rate, nausea, abdominal cramps, hypothermia, vertigo, blurred vision, convulsions and circulatory shock [40]. Bites by this spider are said to be rare and envenomations are mild, and a 2008 study is cited in this account [40] which says that only 2.3% of bites require antivenom. As of 2008, 10 deaths had been reported from Brazilian wandering spider bites and cases of serious envenomation were reported to be rare (0.5%).

Specific toxins have been isolated and identified [40]. A potent component of the venom of *Phoneutria nigriventor* is termed PnTx2-6 (the sixth identified in a series of components). The gene that encodes the production of this protein toxin has been identified and introduced into cultured caterpillar cells *in vitro* with production of toxin. This offers a method of producing the toxin in the laboratory and future studies may result in discoveries with medical benefits from this poisonous agent. Indeed, a report in 2016 concluded that a peptide called antinociception was induced by spider venom component PnPP-19. Antinociception is a synthetic peptide produced as a drug candidate to treat erectile dysfunction [43]. This peptide was designed in the laboratory, based on Brazilian wandering spider venom component PnTx-6. A prior report in 2012 [44] showed that PnTx2-6: "facilitates penile relaxation… through a mechanism [that occurs partially] … *via* increasing nitric oxide… production." Nitric oxide acts as a signal molecule for many processes throughout the body. Nitric oxide was initially studied in 1772 by Joseph Priestley who did early, significant research on the biology of oxygen [45] and a Nobel Prize was awarded in 1998 to Robert F. Furchgott, Louis J. Ignarro, and Ferid Murad for "discoveries concerning nitric oxide as a signaling molecule in the cardiovascular system" [46].

As early as 2010, it was reported that PnTx2-6 slows sodium channels inactivation in nitrergic neurons thus allowing Ca^{2+} influx to facilitate signaling by nitric oxide. This results in increased nitric oxide production and has a relaxing effect [47]. This has relevance to the penile erection effect described earlier. *Phoneutria nigriventor* toxins PnTx2-5 and PnTx-6 were shown to delay the fast inactivation kinetics of neuronal-type sodium channels, and the binding site of this molecules was explored [48]. The DNA sequence responsible for coding for the precursor to toxin PnTx-6 was cloned and sequenced and found to have 403 nucleotides [49]. The 'reason' for the design of such complex and intricate molecules that have such drastic effects on humans, including the specific binding and interference with nerve transmission, is unknown to science.

BROWN RECLUSE SPIDER

The Missouri Department of Conservation website has good information about the brown recluse spider [50]. This spider belongs to the family called *Loxoscelidae* which translated means venomous, six-eyed spiders. The spider is known colloquially as the 'violin spider'. This name is descriptive because of a shape that resembles a violin which is on the back of the head-thorax (not abdomen) of the spider (Fig. **10**). The spider is described as small with a body only about ¼ inch, or up to one inch including spread legs which are darker in color than the body and are long and slim. The color is 'grayish-yellow-brown' and the abdomen is covered with gray hairs [50]. Females are larger than males. The webs made by this spider are 'small, irregular, and untidy' and the spiders apparently do not spend much time on them. The term 'recluse' is appropriate. The spiders hide well and are found in drawers, closets, attics, basements, and behind furniture; they cannot climb smooth un-textured surfaces and are sometimes found trapped in bathtubs or sinks. They are sometimes discovered in garments that were 'packed-away'. The Missouri Conservation site warns that they are rarely seen (reclusive) but if you see one there are probably more. Their diet is small insects and other spiders. They are described as 'running' spiders and rather than using a web to entrap its prey, like the wolf, it chases its prey and uses its venom to subdue it quickly [50]. It is said that "… brown recluses cannot easily bite humans unless they are pressed against our skin… as when they are suddenly trapped between a garment and our bodies, or if they are exploring our bed sheets and we roll on top of them" [50].

Fig. (10). A brown recluse spider with the characteristic violin-shaped darker area just back of the head (not on the abdomen).

Eleven species of the brown recluse spider are indigenous to the continental United States (Fig. **11**) [51]. This website from Pennsylvania University also states that four species are known to be harmful to humans. More specifically with respect to range, brown recluse spiders are said to be 'established' in the following 15 states: Alabama, Arkansas, Georgia, Illinois, Indiana, Iowa, Kansas, Kentucky, Louisiana, Mississippi, Missouri, Ohio, Oklahoma, Tennessee, and Texas. 'Isolated occurrences' have been reported from: Arizona, California, Washington, D.C., Florida, North Carolina, New Jersey, Pennsylvania, Washington, and Wyoming [51].

Fig. (11). Regions of the Continental United States where the brown recluse spider is established.

Clinical Effects of the Venom

Brown recluse spider venom is a complex mixture of components. One study [52], using a proteomics approach for analysis, reported the venom contained 39 different protein components. Proteomics is a branch of biotechnology that applies techniques of molecular biology, biochemistry, and genetics to analyze the structure and function of proteins and to discover the interaction of proteins with genes. Proteomics also organizes such information into databases with the objective of understanding cell functions. The proteins in the venom have a variety of functions. They were speculated to account for the different types of lesions produced by the complex toxin, and to be responsible for certain observed, physiological actions. Also, some component proteins stabilize the toxin molecules. Because these toxins are proteins, identifying the genes that encode the information required for their biosynthesis could lead to molecular biology techniques to express the genes *in vitro* in cell cultures with controlled production of large amounts of pure toxin for research projects.

Researchers have developed a rabbit model to study the development of the lesions produced by the bite of the brown recluse spider [53]. The site of the bite of this spider produces tissue damage that is called an eschar because it resembles the scab (eschar) that typically develops after a significant burn (Fig. **12**). In the rabbit model, both the size and depth of the eschar were related to the dose of the brown recluse spider venom injected into the skin. There was also a dose relationship for the development of factors associated with blood clotting that could be related to the known effects of the toxin to cause disseminated intravascular coagulation – a dreaded complication seen after brown recluse envenomation.

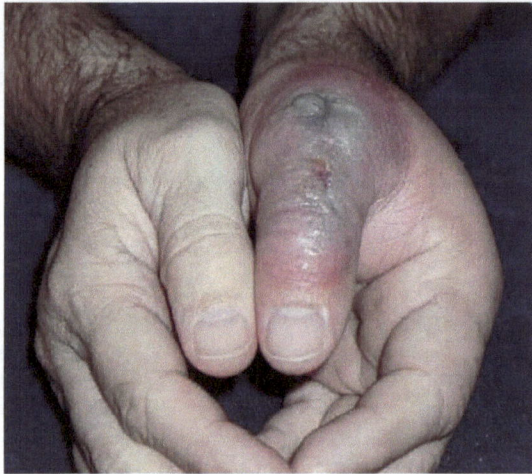

Fig. (12). A necrotizing bite from a brown recluse spider.

In another study [54], tissues after appropriate staining, were studied under the microscope. The tissues were taken from rabbits after manual injection of brown recluse toxin. Brown recluse spider bites cause significant skin and deeper tissue lesions that range from chronic, necrotic ulcers to acute, life-threatening sepsis resulting from bacterial infection. The cited study also reported successful experiments with the use of vacuum-assisted wound closure. The effects observed at the microscopic level helped, in part, to explain the gross effects seen in wounds caused by the toxin. Cells showed a type of mixed-cell infiltration typical of the inflammation, coagulative tissue necrosis, and vasculitis seen in spider bites. Necrosis refers to cell death and vasculitis refers to inflammation of blood vessels (or a single vessel). Based on results of stained tissue they described a type of damage at the wound site, which they characterized as "mummified coagulative necrosis of the epidermis and dermis" (dermis is the thick layer of living tissue below the epidermis which is the outer skin layer). This indicates a serious penetration and pathological involvement of the deep, living skin tissue by the venom. Neutrophils (white blood cells) were recruited to the site, and vasculitis of small blood vessels was apparent, and in some cases large-vessel vasculitis was seen. It appeared that eosinophils recruited into the tissues likely contributed to the pathology at the site [54].

A principal toxic component of brown recluse venom is sphingomyelinase D, an enzyme that degrades sphingomyelin which is a component of most brain and nervous tissue [55]. It is a complex molecule made of phosphoryl derivatives of sphingosine and choline. These workers showed that a glycoprotein, present normally in the human body, called amyloid protein (whose function is poorly understood although the protein is ubiquitous) has a role in the development by the toxin of tissue necrosis. Thus, it is abundantly clear that the venom of the brown recluse spider contains one or more factors that chemically produce skin necrosis. Indeed, as early as 1984, a component was purified from the brown recluse spider venom that caused necrosis of the skin. The purified factor had sphingomyelinase D activity (see above). Specialized laboratory techniques using antibody that inhibited the skin necrosis, with the aid of electron microscopy, were used to identify that the venom attached to the plasma membrane of erythrocytes They speculated that this mechanism underlies the cutaneous inflammatory reaction produced by brown recluse spider venom [56].

In 1979, two former colleagues, J. T. Barrett, and B. Campbell, at the University of Missouri along with others, showed that only one of seven or eight major (plus three or four minor) protein components of the brown recluse spider venom caused necrosis in guinea pig skin [57]. Hyperimmunization of rabbits with venom caused the rabbits to be resistant to formation of lesions by the injected toxin. The spreading activity of the venom in guinea pig skin was inhibited by up to 71% by antivenom. They also reported: "No lesions developed when high concentrations of venom were intradermally injected into the skin of sacrificed guinea pigs, indicating that an interaction of body constituents and venom is essential for the development of a lesion" [57].

Hyperbaric oxygen therapy (a special interest of mine) is reported to be controversial as a treatment of bites by the brown recluse spider [58]. This study reported new findings based on exposing toxin to hyperbaric oxygen (HBO) with the result that HBO-treated toxin venom was not altered.

THE TRANTULA

The tarantula is included here, not because of the toxicity of its venom or threat to humans, which is slight, but because of its size, it is truly awesome in appearance, and because it is the source of classic legends. For many people in the United States, the tarantula is instantly imagined when a venomous spider is mentioned.

The word tarantula is derived from tarantella, a folk dance with one or two couples that originated in Italy. The origin of this dance is connected with tarantism, a disease, or perhaps a form of hysteria, that occurred in Italy during the 15th to 17th centuries. It was obscurely associated with the bite of the tarantula spider and victims seemingly were cured by frenzied dancing. The names for the spider and the dance seem ultimately to be derived from the town of Taranto, Italy as reported by the on-line Encyclopedia Britannica [12] which also recounts that tarantellas were written for the piano by Frédéric Chopin, Franz List and Carl Maria von Weber. So, there is a connection with the feared tarantula and the art of great music.

A recommended site for tarantula biology [59] provides a wealth of information. This spider has accumulated many names: bird-eater, bird spider, tarantula, baboon-spider, monkey-spider, and murderer of horses are some of its, mostly underserved, monikers. This web site begins with an explanation that the name bird-eating spiders originated because of drawings by a German biologist Maria Sibilla Merian [1647-1717] showing the spider with a humming bird which is defending its nest. This reference states that other naturalist have reported evidence that birds can be attacked.

Tarantulas (family *Theraphosida*) (Fig. **13**) exist over a wide range, from 40 degrees north to 40 degrees south latitude (on all continents except Antarctica) and varied terrains from semi-deserts and deserts to moist equatorial forests (swamps excluded) [59]. They are abundantly found is tropical forests of America, Asia, and Africa. Oddly, tarantulas have two pairs of lungs. Difference among tarantulas are said to be primarily because of lifestyle differences [59]. Arboreal tarantulas have longer bodies and legs and can run rapidly. Terrestrial tarantulas are more robust and have shorter legs and they move more slowly.

Fig. (13). The tarantula.

Tarantulas have defensive mechanisms that differ considerably from species to species. All are hair covered (covered with setae). This hair has several functions. Most species found in the Americas (there are exceptions) have thousands of 'urticating' hairs (urticate refers to producing a stinging sensation as by the nettle). These hairs provide protection and when threatened some tarantulas can actually flick hairs by using their rear legs rubbed against their bodies [59]. There are six different types of urticating hairs but they are absent in Old World tarantulas. Other defense mechanisms of tarantulas include an aggressive stance with the front legs in the air with fangs bared. This site states that Old World tarantulas are more aggressive and their venom is more toxic than New World tarantulas. New World tarantulas will, at the moment of danger, turn toward the attacker and briskly rub their hind legs against their opisthosoma throwing the urticating hairs in the direction of the enemy. The cloud of small hairs can get into the mucous membranes of small mammals to cause oedema and difficulty in breathing; the latter can be fatal. In humans this cloud of urticating hairs can cause allergic skin reactions which can manifest "… as inflammation, rash and itching. The reactions can last for several hours or days" [59]. The spider's most powerful defense is the fangs which are capable of producing a painful bite in humans. When threatened they rear up and sometimes fall over on their backs and are even known to be capable of making a hissing sound (produced by body rubbing using a specialized "stridulating apparatus)".

I recall my first experience with a tarantula when I was a small boy in Oklahoma. Toward evening, I encountered a large black and white colored tarantula near the barn on a farm. I was curious and somewhat frightened at the behavior of this large spider which bared its fangs, reared up and subsequently fell over backwards. I had never before encountered such a large spider. I have since learned that all tarantulas have a 'twilight lifestyle'. Relatives living in Denison Texas (which is north of Dallas) reported that their neighborhood was called 'tarantula hill' because of infestation by tarantulas.

It is said: "Generally, tarantula bites are not more serious than the sting of a bee or wasp. Often the bite is a 'dry bite' in which no venom is released." All tarantulas are predators and feed on a variety of insects, small mammals, amphibians, and lizards (and even small snakes). They are cannibalistic and will consume their own young. The tarantula's poison is sufficient to kill its prey, and the toxin works 'almost instantly' [59].

The venom of many species of tarantulas that produce venoms to assist in catching their prey contains protein toxins that act by binding in a very specific way to specific components on nerve cells. They bind specifically to channels or pores that are involved in critical transport of positively charged ions of calcium (and sometimes potassium). The toxin binding is so specific that binding only occurs to specific conformations of the channels and these conformations are important to proper functions of nerves. It will be helpful in understanding the toxicity mechanism of venoms of the tarantula to briefly review some aspects of nerve functioning [60]. Electrical signals travelling along excitable cell membranes activate 'voltage-gated' ion channels. These are structures that have pores which can open or close and do so to varying degrees. This process leads to a cascade of signals that induce various cell functions including: neurotransmitter release (necessary for electrical signals to pass from nerve to nerve or nerve to muscle by a chemically-mediated step), hormone secretion, gene transcription, and other cellular functions. Maladies result from aberrant signaling and these include: pain syndromes, cardiac arrhythmias, and paralysis. More than 80 voltage-gated ion channels are known [60]. All are made of proteins and information for their biosyntheses is encoded in the genome, transcribed into RNA, and translated into proteins. It continues to be challenging to understand the role and relationships of these channels. Some important tarantula venoms are toxic *via* mechanisms that interfere with nerve transmission at specific ion channels.

One important venom known as 'cystine knot toxin' (also called JZTx-27), from the tarantula *Chilobrachys jingzhao,* was found (oddly) to stabilize the sodium channel in bacteria when in the deactivated state! This research [12] is reported here because of its future importance for applications designed to learn the specific nature of ion-channels in nerve cells, an enormously difficult problem. These poisons also provide great hope for understanding diseases of the CNS, and new therapies for treatment.

Vanillotoxins are components of the venom of various tarantulas. They are poly-peptides (small proteins) and one that has been much studied is Vanillotoxin-1. It is also referred to as Pc1a and VaTx1. All possess a specific structure and are called 'cystine knot' toxins because of the molecular arrangement resulting from a particular sequence of an amino acid known as cystine. Cystine contains sulfur as a disulfide bridge and in proteins this results in a particular shape in three dimensions (Fig. **14**). There are three subtypes of these toxins and they specifically block cation channels in neuronal cells. The mechanisms by which these toxins have effects on humans is actively being researched. For example, the venom of the tarantula *P. Cambridgei* is known to produce significant pain. The amount of this toxin in a bite from this species is too small to otherwise be a serious health threat to humans [61]. These toxins bind to extracellular pore domains of peripheral nerves causing pores to open with influx of calcium cations which activates the pain response [61]. Study of these toxins may provide new means of pain management, a much needed objective today, when so many are suffering from an opioid epidemic that is producing deaths from the addicting side-effects of many currently prescribed pain medications (see Chapter 9).

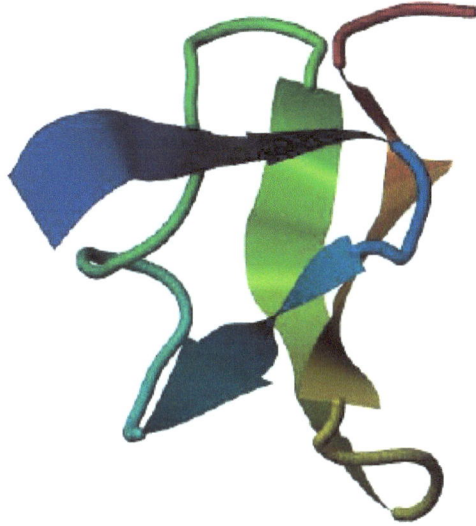

Fig. (14). Vanillotoxin as a classical ribbon diagram to reveal the complex shape of the venom.

Omega-grammatoxin (Fig. **15**) is another significant toxin affecting humans from tarantulas [62]. It has been isolated from *Grammostola spatulata*. This venom inhibits P-, Q-, and N-type voltage-gated calcium channels in neurons. Its action is complex but laboratory studies have added to understanding of the biological mechanisms for control of transmission of electrical signals by neurons. This toxin has high affinity for pores when they are closed but low affinity when pores are open (activated). Therefore, the toxin preferentially binds to closed channels and it binds in a region that contains the voltage-sensing domain of the cell pore. This results in blocking of normal depolarization which is necessary for normal nerve function. Grammotoxin also binds with lower affinity to potassium channels with further toxic results [62]. Sites termed 'low-voltage-activated' calcium channels are specific sites of binding and physiological action of tarantula toxins [63]. This is a rare site for toxin action and details of the mechanisms of action involve complex binding reactions that are not yet fully understood [63]. However, to understand the toxic mechanism of tarantula venoms it is useful to describe their mechanism of action in more detail. Voltage-gated calcium channels are membrane proteins that assist cells in certain functions that may best be understood broadly as 'cellular communications'. This occurs *via* both electrical and chemical signaling and it occurs in a wide variety of cells and organisms, including, of course, in humans. Ca^{2+} enters these channels but also serves as a 'second messenger' inside the cell to initiate and control many necessary cellular functions. These functions include muscle contraction, hormone secretion, neurotransmitter release, and regulation of gene expression as previously stated. Ca^{2+} channels are made of four domains (proteins) each with complex structures that are functionally and pharmacologically distinct [63].

Fig. (15). Grammotoxin, showing the 'cystine knot', a structure found in several tarantula toxins.

Studies of these venoms have contributed to the biochemistry and physiology of sending and receiving of signals, and initiation of cellular actions. This offers great promise for developing therapeutics. One exciting area of current study, reported in 2016, involves the neuroscience of pain as reported in a website of the University of California at San Francisco [64]. This site describes a report of a study published in *Nature* by lead researcher David Julius. It describes two toxins isolated from the venom of *Heteroscodra maculate,* a large West African tarantula (described to be as large as one's hand). It is locally known descriptively as the 'ornamental baboon' and 'Togo starburst'. With massive fangs it can envenomate to cause excruciating pain. The mechanisms described in this chapter are involved in the production of pain; however, a unique type of sodium channel is primarily responsible. This report suggests that in addition to its applications for pain research, this toxin by its manipulating of a specific sodium channel may have medical applications for epilepsy, autism and Alzheimer's disease [64]. Another site describes research using venom from the Peruvian green velvet tarantula [65]. A group of researchers from the University of Queensland in Brisbane, Australia have found that a peptide toxin (ProTx-II) in venom from this tarantula (*Thrixopelma pruriens*) has high and selective biological activity that targets pain sensation receptors. Thus, in research to discover mechanisms of pain production, there is hope to find pain-killing drugs without addicting side-effects.

NOTES

[1] Elwyn Brooks White. Letter to Carrie A. Wilson, May 1, 1951. Bartlett's Familiar Quotations. John Bartlett. Edited by Emily Morison Beck and the editorial staff of Little, Brown and Company, 15th edition, 1980.

[2] An anonymous Nursery Rhyme. *Mother Goose Tales* originated with Charles Perrault [1628 – 1703] who published *Contes de Ma Mèrel'Oye* [1697], a collection of traditional tales. John Newbury [1713-1767], who originated the publication of children's books, first published the rhyme in1765. Bartlett's Familiar Quotations, pp. 929 – 932. John Bartlett, Edited by Emily Morison Beck, Little, Brown and Company, 15th ed., 1980.

[3] The Bible, 1 Samuel, 24.

REFERENCES

[1]   Spider Web Facts: Silk Properties and Purpose of Different Webs of Spiders [Internet]. [cited 2017 Nov 17]. Available from: http://www.pestproducts.com/spider-webs.htm

[2]   Römer L, Scheibel T. The elaborate structure of spider silk: structure and function of a natural high performance fiber. Prion [Internet] Taylor & Francis , 2008 [cited 2017 Nov 17];2(4): 154-61. Available from: http://www.ncbi.nlm.nih.gov/pubmed/19221522 [http://dx.doi.org/10.4161/pri.2.4.7490]

[3]   Agnarsson I, Kuntner M, Blackledge TA. Bioprospecting finds the toughest biological material: extraordinary silk from a giant riverine orb spider. PLoS One [Internet] Public Library of Science , 2010 Sep 16; [cited 2017 Nov 17];5(9): e11234. Available from: http://www.ncbi.nlm.nih.gov/pubmed/20856804

[4]   Myths and Legends of the Sioux by Marie L. McLaughlin, 1990 | Online Research Library: Questia [Internet]. [cited 2017 Nov 19]. Available from: https://www.questia.com/library/124977/myths-an--legends-of-the-sioux

[5]   Mark P, Morford R. Classical Mythology. U.S.: Oxford University Press 1999.

[6]   Charlotte's Web Book Review [Internet]. [cited 2017 Nov 19]. Available from: https://www.commonsensemedia.org/book-reviews/charlottes-web

[7]   The Fable of the Spider and the Fly (1923) - IMDb [Internet]. [cited 2017 Nov 19]. Available from: http://www.imdb.com/title/tt0146710/

[8]   A Noiseless Patient Spider | Representative Poetry Online [Internet]. [cited 2017 Nov 19]. Available from: https://rpo.library.utoronto.ca/poems/noiseless-patient-spider

[9]   Spider Legends at Spiderzrule - the best site in the world about spiders, redbacks, huntsmen, garden orb weaver, funnel web, black widow, recluse, hobo spider, daddy long legs. [Internet]. [cited 2017 Nov 19]. Available from: http://spiderzrule.com/legends.htm

[10] Arachne [Internet]. [cited 2017 Nov 18]. Available from: https://www.greekmythology.com/Myths/Mortals/Arachne/arachne.html

[11] Leaflor L. The Symbolic Spider That Wove Its Way Through History. [Internet]. [cited 2017 Nov 18]. Available from: http://www.ancient-origins.net/myths-legends/symbolic-spider-wove-its-way-through-history-002215?nopaging=1

[12]   Tang C, Zhou X, Nguyen PT, Zhang Y, Hu Z, Zhang C, *et al.* A novel tarantula toxin stabilizes the deactivated voltage sensor of bacterial sodium channel. FASEB J [Internet] FASEB , 2017 Jul 1; [cited 2017 Nov 17];31(7): 3167-78. Available from: http://www.ncbi.nlm.nih.gov/pubmed/28400471 [http://dx.doi.org/10.1096/fj.201600882R]

[13]   Garb JE, Hayashi CY. Molecular Evolution of α-Latrotoxin, the Exceptionally Potent Vertebrate Neurotoxin in Black Widow Spider Venom. Mol Biol Evol [Internet] Oxford University Press , 2013 May 1; [cited 2017 Nov 17];30(5): 999-1014. Available from: https://academic.oup.com/mbe/article-lookup/doi/10.1093/molbev/mst011 [http://dx.doi.org/10.1093/molbev/mst011]

[14]   Black Widow Spider Facts [Internet]. [cited 2017 Nov 17]. Available from:https:// www.livescience.com/39919-black-widow-spiders.html

[15]   Yan Y, Li J, Zhang Y, Peng X, Guo T, Wang J, *et al.* Physiological and biochemical characterization of egg extract of black widow spiders to uncover molecular basis of egg toxicity. Biol Res [Internet] BioMed Central , 2014 May 16; [cited 2017 Nov 17];47(1): 17. Available from: http://www.biolres.com/content/47/1/17 [http://dx.doi.org/10.1186/0717-6287-47-17]

[16]   Goel SC, Yabrodi M, Fortenberry J. Recognition and successful treatment of priapism and suspected black widow spider bite with antivenin. Pediatr Emerg Care 2014; 30(10): 723-4. [http://dx.doi.org/10.1097/PEC.0000000000000235] [PMID: 25275351]

[17]   Fact Sheet OSHA. OSHA Fact Sheet, Black Wido Spider [Internet]. [cited 2017 Nov 17]. Available from: https://www.osha.gov/OshDoc/data_Hurricane_Facts/black_widow_spider.pdf

[18]   Yan S, Wang X. Recent Advances in Research on Widow Spider Venoms and Toxins. Toxins (Basel) 2015; 7(12): 5055-67. [http://dx.doi.org/10.3390/toxins7124862] [PMID: 26633495]

[19]   Representative LD. Representative LD 50 Values [Internet]. [cited 2017 Jan 1]. Available from: http://biology.unm.edu/toolson/biotox/representative_LD50_values.pdf

[20]   Khvotchev MV. Latrotoxin Encyclopedia of Neuroscience [Internet]. , Elsevier 2009; pp. [cited 2017 Nov 17];391-4. Available from: http://linkinghub.elsevier.com/retrieve/pii/B9780080450469013826 [http://dx.doi.org/10.1016/B978-008045046-9.01382-6]

[21]   Grishin E. Black widow spider toxins: the present and the future. Toxicon [Internet] Pergamon , 1998 Nov 1; [cited 2017 Nov 17];36(11): 1693-701. Available from: http://www.sciencedirect.com/ science/article/pii/S0041010198001627?via%3Dihub [http://dx.doi.org/10.1016/S0041-0101(98)00162-7]

[22]   Ushkaryov YA, Rohou A, Sugita S. alpha-Latrotoxin and its receptors. Handb Exp Pharmacol [Internet] Europe PMC Funders , 2008 [cited 2017 Nov 17];(184): 171-206. Available from: http://www.ncbi.nlm.nih.gov/pubmed/18064415

[23]   Südhof TC. α-Latrotoxin and Its Receptors: Neurexins and CIRL/Latrophilins. Annu Rev Neurosci [Internet] Annual Reviews 4139 El Camino Way, PO Box 10139, Palo Alto, CA 94303-0139, USA , 2001 Mar 28; [cited 2017 Nov 17];24(1): 933-62. Available from: http://www.annualreviews.org/ doi/10.1146/annurev.neuro.24.1.933

[24]   Pineda SS, Wilson D, Mattick JS, King GF. The lethal toxin from Australian funnel-web spiders is encoded by an intronless gene. PLoS One [Internet] Public Library of Science , 2012 [cited 2017 Nov 19];7(8): e43699. Available from: http://www.ncbi.nlm.nih.gov/pubmed/22928020

[25]   Funnel-web Spiders - Australian Museum [Internet]. [cited 2017 Nov 19]. Available from: https://australianmuseum.net.au/funnel-web-spiders-group

[26]   Browne GJ. Near fatal envenomation from the funnel-web spider in an infant. Pediatr Emerg Care [Internet] , 1997 Aug; [cited 2017 Nov 19];13(4): 271-3. Available from: http://www.ncbi.nlm. nih.gov/pubmed/9291517

[http://dx.doi.org/10.1097/00006565-199708000-00009]

[27]   Bites by Australian mygalomorph spiders (Araneae, Mygalomorphae), including funnel-web spiders (Atracinae) and mouse spiders (Actinopodidae: Missulena spp). Toxicon [Internet] Pergamon , 2004 Feb 1; [cited 2017 Nov 19];43(2): 133-40. Available from: https://www.sciencedirect.com/ science/article/pii/S0041010103003386

[28]   Isbister GK, Warner G. Acute myocardial injury caused by Sydney funnel-web spider (Atrax robustus) envenoming. Anaesth Intensive Care 2003; 31(6): 672-4.
[PMID: 14719431]

[29]   Miller MK, Whyte IM, White J, Keir PM. Clinical features and management of Hadronyche envenomation in man. Toxicon [Internet] , 2000 Mar; [cited 2017 Nov 19];38(3): 409-27. Available from: http://linkinghub.elsevier.com/retrieve/pii/S0041010199001713
[http://dx.doi.org/10.1016/S0041-0101(99)00171-3]

[30]   Del Brutto OH. Neurological effects of venomous bites and stings: snakes, spiders, and scorpions. Handb Clin Neurol 2013; 114: 349-68. [Review].
[http://dx.doi.org/10.1016/B978-0-444-53490-3.00028-5] [PMID: 23829924]

[31]   Nicholson GM, Little MJ, Birinyi-Strachan LC. Structure and function of δ-atracotoxins: lethal neurotoxins targeting the voltage-gated sodium channel. Toxicon [Internet] , 2004 Apr; [cited 2017 Nov 19];43(5): 587-99. Available from: http://linkinghub.elsevier.com/retrieve/pii/S0041010104 000613
[http://dx.doi.org/10.1016/j.toxicon.2004.02.006]

[32]   Nicholson GM, Graudins A. Spiders of medical importance in the Asia-Pacific: Atracotoxin, latrotoxin and related spider neurotoxins. Clin Exp Pharmacol Physiol [Internet] Blackwell Science Pty , 2002 Sep 1; [cited 2017 Nov 19];29(9): 785-94. Available from: http://doi.wiley.com/ 10.1046/j.1440-1681.2002.03741.x

[33]   Hodgson WC. Pharmacological Action of Australian Animal Venoms. Clin Exp Pharmacol Physiol [Internet] , 1997 Jan 1; [cited 2017 Nov 19];24(1): 10-7. Available from: http://doi.wiley.com/ 10.1111/j.1440-1681.1997.tb01776.x
[http://dx.doi.org/10.1111/j.1440-1681.1997.tb01776.x]

[34]   Agatoxins: ion channel specific toxins from the american funnel web spider, Agelenopsis aperta Toxicon [Internet] Pergamon , 2004 Apr 1; [cited 2017 Nov 19];43(5): 509-25. Available from: https://www.sciencedirect.com/science/article/pii/S004101010400056X

[35]   The ω-atracotoxins: Selective blockers of insect M-LVA and HVA calcium channels Biochem Pharmacol [Internet] Elsevier , 2007 Aug 15; [cited 2017 Nov 19];74(4): 623-38. Available from: http://www.sciencedirect.com/science/article/pii/S0006295207003279?via%3Dihub

[36]   Nicholson GM, Walsh R, Little MJ, Tyler MI. Characterisation of the effects of robustoxin, the lethal neurotoxin from the Sydney funnel-web spider Atrax robustus, on sodium channel activation and inactivation. Pflugers Arch Eur J Physiol [Internet] Springer-Verlag , 1998 Apr 27; [cited 2017 Nov 19];436(1): 117-26. Available from: http://link.springer.com/10.1007/s004240050612
[http://dx.doi.org/10.1007/s004240050612]

[37]   Alphabet of Ben Sira [Internet]. [cited 2017 Nov 19]. Available from: https:// www.jewishvirtuallibrary.org/alphabet-of-ben-sira

[38]   Brazilian Wandering Spiders: Bites and Other Facts [Internet]. [cited 2017 Nov 20]. Available from: https://www.livescience.com/41591-brazilian-wandering-spiders.html

[39]   Representative LD-50 Values [Internet]. [cited 2017 Nov 20]. Available from: http://biology.unm.edu/ toolson/biotox/representative_LD50_values.pdf

[40]   Venom from Brazilian Wandering Spider slated as next Viagra [Internet]. [cited 2017 Nov 20]. Available from: http://www.ibtimes.co.uk/venom-brazilian-wandering-spider-slated-next- viagra-1488235

[41] wandering spider | arachnid | Britannica.com [Internet]. [cited 2017 Nov 20]. Available from: https://www.britannica.com/animal/wandering-spider

[42] Brazilian Wandering Spider - Facts, Bite & Habitat Information [Internet]. [cited 2017 Nov 20]. Available from: https://animalcorner.co.uk/animals/brazilian-wandering-spider/

[43] Freitas A, Pacheco D, Machado M, Carmona A, Duarte I, DeLima M. PnPP-19, a spider toxin peptide, induces peripheral antinociception through opoid and cannabinoid receptors and inhibition of neutral endopeptidase. Br J Pharmacol [Internet] , 2016 May; [cited 2017 Nov 20];173(9): 1491-501. Available from: http://doi.wiley.com/10.1111/bph.13448

[44] Nunes KP, Wynne BM, Cordeiro MN, Borges MH, Richardson M, Leite R, *et al.* Increased cavernosal relaxation by Phoneutria nigriventer toxin, PnTx2-6, *via* activation at NO/cGMP signaling. Int J Impot Res [Internet] Nature Publishing Group , 2012 Mar 1; [cited 2017 Nov 20];24(2): 69-76. Available from: http://www.ncbi.nlm.nih.gov/pubmed/21975567

[45] Brown OR. Oxygen, the breath of life: boon and bane in human health, disease, and therapy. Bentham Science 2017; pp. 56-61p.
[http://dx.doi.org/10.2174/97816810842511170101]

[46] Physiology or Medicine for 1998 - Press Release [Internet]. [cited 2017 Nov 20]. Available from: https://www.nobelprize.org/nobel_prizes/medicine/laureates/1998/press.html

[47] Nitric Oxide-Induced Vasorelaxation in Response to PnTx2-6 Toxin from Phoneutria nigriventer Spider in Rat Cavernosal Tissue. J Sex Med [Internet] Elsevier , 2010 Dec 1; [cited 2017 Nov 20];7(12): 3879-88. Available from: http://www.sciencedirect.com/science/article/pii/S1743609515327995

[48] Matavel A, Fleury C, Oliveira LC, Molina F, de Lima ME, Cruz JS, *et al.* Structure and Activity Analysis of Two Spider Toxins That Alter Sodium Channel Inactivation Kinetics †. Biochemistry [Internet] , 2009 Apr 14; [cited 2017 Nov 20];48(14): 3078-88. Available from: http://pubs.acs.org/doi/abs/10.1021/bi802158p

[49] Electrophysiological characterization and molecular identification of the Phoneutria nigriventer peptide toxin PnTx2-6 FEBS Lett [Internet] No longer published by Elsevier , 2002 Jul 17; [cited 2017 Nov 20];523(1-3): 219-3. Available from: http://www.sciencedirect.com/science/article/pii/S0014579302029885

[50] Brown Recluse (Violin Spider) | MDC Discover Nature [Internet]. [cited 2017 Nov 20]. Available from: https://nature.mdc.mo.gov/discover-nature/field-guide/brown-recluse-violin-spider

[51] Brown Recluse Spiders — Department of Entomology — Penn State University [Internet]. [cited 2017 Nov 20]. Available from: http://ento.psu.edu/extension/factsheets/brown-recluse-spiders

[52] dos Santos L, Dias N, Roberto J, Pinto A, Palma M. Brown Recluse Spider Venom: Proteomic Analysis and Proposal of a Putative Mechanism of Action Protein Pept Lett [Internet] , 2009 Aug 1; [cited 2017 Nov 20];16(8): 933-43. Available from: http://www.eurekaselect.com/openurl/content.php?genre=article&issn=0929-8665&volume=16&issue=8&spage=933

[53] McGlasson DL, Harroff HH, Sutton J, Dick E, Elston DM. Cutaneous and systemic effects of varying doses of brown recluse spider venom in a rabbit model. Clin Lab Sci 2007; 20(2): 99-105.
[PMID: 17557708]

[54] Elston DM, Eggers JS, Schmidt WE, Storrow AB, Doe RH, McGlasson D, *et al.* Histological Findings After Brown Recluse Spider Envenomation Am J Dermatopathol [Internet] , 2000 Jun; [cited 2017 Nov 20];22(3): 242-6. Available from: http://content.wkhealth.com/linkback/ openurl?sid=WKPTLP:landingpage&an=00000372-200006000-00006
[http://dx.doi.org/10.1097/00000372-200006000-00006]

[55] Gates C, Rees R. Serum amyloid P component: its role in platelet activation stimulated by sphingomyelinase D purified from the venom of the brown recluse spider (Loxosceles reclusa) Toxicon [Internet] Pergamon , 1990 Jan 1; [cited 2017 Nov 20];28(11): 1303-5. Available from:

http://www.sciencedirect.com/science/article/pii/004101 019090095O

[56] Rees RS, Nanney LB, Yates RA, King LE. Interaction of brown recluse spider venom on cell membranes: the inciting mechanism? J Invest Dermatol [Internet] , 1984 [cited 2017 Nov 20];83(4): 270-5. Available from: http://www.ncbi.nlm.nih.gov/pubmed/6481179

[57] Barrett JT, Gebel HM, Elgert KD, Campbell BJ, Finke JH. Inactivation of Complement by Loxosceles Reclusa Spider Venom. Am J Trop Med Hyg [Internet] , 1979 Jul 1; [cited 2017 Nov 20];28(4): 756-62. Available from: http://www.ajtmh.org/content/journals/10.4269/ajtmh.1979.28.756

[58] Merchant ML, Geren CR, Hinton JF. Effect of Hyperbaric Oxygen on Sphingomyelinase D Activity of Brown Recluse Spider (Loxosceles reclusa) Venom as Studied by 31P Nuclear Magnetic Resonance Spectroscopy. Am J Trop Med Hyg [Internet] , 1997 Mar 1; [cited 2017 Nov 20];56(3): 335-8. Available from: http://www.ajtmh.org/content/journals/10.4269/ajtmh.1997.56.335

[59] Tarantula biology | Theraphosids (tarantulas) of the World. Keeping and breeding in captivity [Internet]. [cited 2017 Nov 21]. Available from: http://tarantulas.su/en/biology

[60] Tilley DC, Eum KS, Fletcher-Taylor S, Austin DC, Dupré C, Patrón LA, *et al.* Chemoselective tarantula toxins report voltage activation of wild-type ion channels in live cells. Proc Natl Acad Sci U S A [Internet] National Academy of Sciences , 2014 Nov 4; [cited 2017 Nov 21];111(44): E4789-96. Available from: http://www.ncbi.nlm.nih.gov/pubmed/25331865 [http://dx.doi.org/10.1073/pnas.1406876111]

[61] Receptor-targeting mechanisms of pain-causing toxins: How ow? Toxicon [Internet]. Pergamon , 2012 Sep 1; [cited 2017 Nov 21];60(3): 254-64. Available from: http://www.sciencedirect.com/science/article/pii/S0041010112004424?via%3Dihub

[62] McDonough SI, Lampe RA, Keith RA, Bean BP. Voltage-dependent inhibition of N- and P-type calcium channels by the peptide toxin omega-grammotoxin-SIA. Mol Pharmacol [Internet] , 1997 Dec; [cited 2017 Nov 21];52(6): 1095-4. Available from: http://www.ncbi.nlm.nih.gov/pubmed/9415720

[63] Salari A, Vega BS, Milescu LS, Milescu M. Molecular Interactions between Tarantula Toxins and Low-Voltage-Activated Calcium Channels. Sci Rep [Internet] Nature Publishing Group , 2016 Jul 5; [cited 2017 Nov 22];6(1): 23894. Available from: http://www.nature.com/articles/srep23894 [http://dx.doi.org/10.1038/srep23894]

[64] Weiler N. Tarantula Toxins Offer Key Insights Into Neuroscience of Pain | UC San Francisco [Internet]. [cited 2017 Nov 22]. Available from: https://www.ucsf.edu/news/2016/06/403166/tarantula-toxins-offer-key-insights-neuroscience-pain

[65] Tarantula toxins converted to painkillers [Internet]. Science Daily , 2016 [cited 2017 Nov 22]; Available from: https://www.sciencedaily.com/releases/2016/02/160229082005.htm

CHAPTER 8

Snakes and their Venoms

Abstract: Snakes, from historical times to the present, have had a bad reputation. The story of the fall in the Garden of Eden has a snake as the betrayer and inducer of original sin. Much of their current reputation is well earned. They are an example of a bearer of poisons that congers up all that we fear about poisoning, and more. As examples of a life form snakes have many unusual features. They lack arms and legs, hands and feet, but they manage to move about, to mate, and to find prey as food. This last function is one where venoms come in handy, given the other deficiencies. There are an amazing variety of poisonous molecules that have been discovered in snake venom. Undoubtedly there are many more to be discovered, and based on what we already know, some of these may be excellent therapeutics for a variety of pathologies. The poison side of these molecules is fascinating with intricate complexities. The toxins are exquisitely specific in the manner in which each can bind and disrupt an essential function – as do the neurotoxins – in a species so far removed from their own lineage as is the mouse or the human. This speaks, at least to me, of the design seen throughout nature.

"Four Snakes gliding up and down a hollow for no purpose that I could see – not to eat, not for love, but only gliding."[1]

Keywords: Alpha-elapitoxin, Antivenin, Black Mamba, Brown Snake, Cobra, Copperhead, Cotton Mouth Moccasin, Eastern Diamondback, Green Mamba, Inland Taipan, King Cobra, Krait, Mamba, Myoneural Junction, Pit vipers Postsynaptic Cobra Neurotoxin, Pygmy Rattlesnake, Timber Rattlesnake, Venom, Western Diamondback.

INTRODUCTION

Snakes never elicit neutral images for most people. From personal experience, most people's impression of snakes runs from fear to revulsion. We notice those few souls who consider them beautiful, or who choose them as pets. In the time and place of my childhood, a snake was something to be warned about in its absence, or killed when it appeared. As a small boy who loved to roam our pastures covered with tall grass and crossed by shallow ravines, the appearance of a blue racer which arched its body in a flashing retreat was scary. An encounter with a 'cotton mouth' moccasin that dropped from a low tree branch into the water just missing falling into my boat was only made more frightening when the snake disappeared underwater and under my boat.

Snakes deserve their own chapter in this book because of the unusually potent venom that some possess together with the means to deliver it (Fig. **1**). Venoms are complex mixtures of chemicals and the most poisonous of snake produce uniquely toxic substances which shall be described in context with particular snakes. Venom is produced in an organ called the venom gland. This gland has an associated powerful muscle which is under the snake control, and the quantity of venom injected by a bite can be adjusted including no venom at all, called a 'dry' bite. The venom passes through channels (ducts) and eventually down hollow fangs which are sharp and generally are curved. A rattlesnake fang is somewhat like a curved hypodermic needle. "Soft tissue around the end of the venom duct and base of the fang seals it from leaking venom… Large venom glands at the base of the jaws are responsible for the distinctively triangular shape of the head… Each fang has a series of seven developing fangs behind the functional fang… large rattlers can have fangs… 4-6 inches long" [1].

Reptiles poorly tolerate extreme heat or cold and seek shelter in underground burrows or under rocks. "In the fall they congregate in rock slides or crevices for their winter hibernation in dens that may shelter hundreds of individuals of several different species… Rattlesnakes give birth to young that develop from eggs retained inside the mother… Newborn rattlesnakes have functioning fangs and venom glands. Their venom is more potent but of lesser quantity than that of their mother… The newborn babies are also equipped with a single button on the end of the tail… The rattle in adults, presumably a warning device, is composed of horny, loosely connected hollow segments, one of which is added every time the snake sheds its skin… rattlesnakes usually shed three or four times a year" [1].

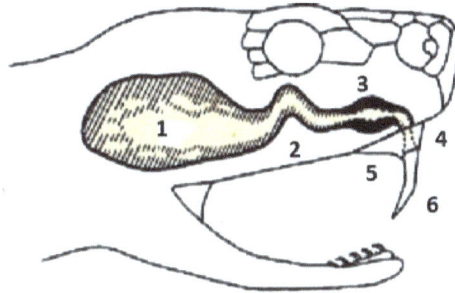

Fig. (1). Snakes have an intricate means of delivering venom: venom gland with compressor muscle (1), primary venom duct (2), accessory gland (3), secondary venom duct (4), fang sheath (5), and fang (6).

The specific snakes to be described are chosen to include those known to have the most toxic venom, or the most notorious reputations which is not the same thing.

Also, I have focused more on snakes found in the United States. This chapter is not a compendium – there are approximately 3,600 species of snakes and most are non-venomous or have venom that is not dangerous to an otherwise healthy adult. There are many websites about snakes; most have a mixture of sensational accounts and verified facts, and many have a '10 most' list – 'most poisonous', 'most deadly', most dangerous.' I have done my best to present verifiable information, to note when legends are being repeated, and to indicate what is only my opinion. This includes conclusions about the comparative toxicity of snake venoms. Obviously, any statement about the effects of venom of any kind, including snake venom, is not the product of the careful laboratory testing that would generate a statistical quantification of venom potency. In some cases, venoms have been tested in the laboratory but they are always tested on some laboratory animal, and not on humans except in rare (mostly accidental) cases. More was written in Chapter 1 about the complication of assessing toxicity for humans, which is generally the focal interest.

Let us dispense with the attribution of 'most dangerous' snake. The criteria must first be established. Are we ranking because of the number of people actually killed, the toxicity of the venom on a volume basis, the toxicity of the most venomous component on a molecular basis, or on other bases? I believe it is sensationalism, but perhaps justifiable to elicit interest, to categorize snakes in all the indicated ways. However, we approach science more closely as we limit the categorization to the comparison of the most toxic component in the venom on a molecular basis; just my opinion, so you will know.

On a molecular basis, the sea snake (*Hydrophis belcheri)* has venom that some regard as the most toxic, more toxic than the venom of the inland taipan *(Oxyuranus microlepidotus)* the land snake with the most toxic venom on a molecular basis. A description of the venom of the inland taipan snake is included in Chapter 1 and more will not be included here. This leaves the option to describe other snakes that are less venomous but which have other interesting characteristics.

RATTLESNAKES

Rattlesnakes are a good place to begin. Of all snakes in the U.S., they probably come to mind first. There are 33 species of venomous snakes characterized as having a segmented rattle at the end of the tail [1]. They are part of the group known as New World vipers, and rattlesnakes are found from southern Canada to central Argentina, but are "… most abundant and diverse in the deserts of the southwestern United States and northern Mexico. Adults usually vary in length from 1.6 to 6.6 feet, but can grow to 8.2 feet. A few species are marked with transverse bands, but most rattlesnakes are blotched with dark diamonds, hexagons, or rhombuses on a lighter background, usually gray or light brown; some are various shades of orange, pink, red, or green" [1, 2].

Fig. (2). Timber rattlesnake.

Timber rattlesnakes (*Crotalus horridus*) are said to be among the most common species in North America and they inhabit the eastern States (Fig. **2**). Also common is *C. viridis* of the western prairies, and the eastern and western diamondbacks, respectively (*C. adamanteus*, and *C. atrox*) [1] (Fig. **3**). There also is the sidewinder, a smaller desert species, the massasauga (*S. catenatus*) and the pygmy rattlesnake (*S. Miliarius*) [1, 2]. The Colorado desert sidewinder, found only in the desert of this State grows only to about 2 feet in length and has 'horns' over its eyes that: "aid in keeping the eyes from being scratched while burrowing" [2]. The most dangerous species according to *The Encyclopeaedia Britannica* are the Mexican west coast rattlesnake *(C. basiliscus)*, the South American rattlesnake (*C. durissus*), and the Mojave rattlesnake (*C. scutulatus*) [1]. This site also says that rattlesnakes are not aggressive and do not attack humans unless provoked; however, it also says: "A rattlesnake bite is very painful, and that of a snake more than …3.3 feet long can be fatal" [1]. Dry bites do occur, and antivenom should not be used in such cases because a significant number of people develop serious allergic reactions to horse serum present in the antivenom. Rattlesnakes are pit vipers, so-called because they have small heat-sensing organs between each eye and nostril which is sensitive enough to detect the heat of a small mammal and is useful in hunting prey. This organ "… provides the snake with stereoscopic heat 'vision' enabling them to detect and accurately strike a living target in complete darkness. Most rattlesnakes live in arid habitats and are nocturnal, hiding during the day but emerging in the evening or at twilight to hunt for prey, which consists primarily of small mammals, especially rodents" [1].

Fig. (3). Eastern diamondback.

The San Diego Zoo site [2] promotes the cause of the rattlesnake by saying: "… we should learn to appreciate the rattlesnake as one of the most efficient and specialized predators on Earth. Amazingly, rattlesnakes are no match for non-venomous king snakes, which are highly resistant to rattlesnake venom; rattlers are a common food item for king snakes."

Centennial, a novel by James A. Michener [3], written for the United States Bicentennial celebration of 1976, is more than 1,000 pages and includes a geologic description of the earth that leads to an unusual description of a rattlesnake that draws the reader into the time and place of an early scene in the novel. The reader may empathize with this rattlesnake. My recollection, having read the story almost 40 years ago, is that the relevant scene is told from the perspective of the rattlesnake. He is gigantic and feels well satisfied that he possesses the terrain at the site where he is coiled. There is also a 'rattlesnake' scene in the movie *True Grit* where the rattlesnake is in no wise empathetic. Rooster Cogburn (played by John Wayne) saves Mattie Ross (played by Kim Darby) after she is cast into a pit with a den of rattlesnakes and is bitten. The scene with the pit of rattlesnakes is dramatic and threatening. It begins to build from the point when the viewer becomes aware of the writhing snakes and progresses with the threat that Mattie would end up in the pit, to the climax when she actually is pushed and falls into the pit and is bitten. Rooster takes Mattie to a doctor on his horse, Bo, who is ridden until he collapses in saving her. The rattlesnake scene is memorable and horrifying. The movie was based on the book *True Grit* written by Charles Portis in 1968 [4].

There is also the book titled *Rattler* which is about a man killing a rattlesnake [5]. Although a man tells the story, it is told from the perspective of the snake. "The snake is seen as human-like and deserving of admiration and respect from the reader… The reader sees the snake as an equal opponent to the man in the story, because the man sees him as such… there is a greater purpose in killing the rattler, that the person sees his defeat of the rattler as undesirable but necessary." [5]. I recall an incident with myself as a small boy and with my dog. One of us, I don't recall whether it was me or my dog, discovered a large snake that was attempting to enter our chicken house. It coiled defensively and raised his head and bared its fangs. I had a mixture of emotions. I was protective of our chickens; I feared for my dog and even my safety because I felt the snake threatened us. I was reasonably sure the snake was not venomous but he was large, about 5 feet long as I later was able to determine. My dog was very protective of me and he, I believe, would have attached a dinosaur if it threatened me. I ran to our house to get a rifle which I never had fired. In the excitement, I was unable to cock the rifle and, instead, shot the snake with my sling shot. I practiced with the sling shot continually and was able to strike the snake just below the head with the first shot. As soon as this happened by dog lunged at the snake and grabbed it. A struggle ensued and the snake lost the battle.

My feelings were similar to the feelings described in the story *Rattler* although my snake was probably a rat snake and was not after our chickens but rats which liked the chicken feed we provided for the chickens. I have no insight into the thoughts of the snake other than it was prepared to defend itself and apparently felt quite capable of doing so.

Rattlesnake Venom

The venom of the rattlesnake varies with the species of the snake, with the size and age of the snake, and with how recently it has expended venom. Venom of all rattlesnakes is a complex mixture of enzymes and protein toxins with a variety of biological actions.

Rattlesnake venoms all contain hemotoxic components. These toxins damage tissues, affect circulation of blood to cause hemorrhaging, and destroy red blood cells. In addition, the venom of some rattlesnakes (those which are most toxic for humans) also contain neurotoxins. Rattlesnake venoms with dominant neurotoxic properties are the venoms of baby rattlesnakes of various species, and the venom of the Mojave rattler [6]. These poisons are related in their toxic actions to that of the neurotoxic venoms of the sea snake, the coral snake, and cobras which have venoms with neurotoxins as the dominate toxin.

Rattlesnake neurotoxic venom has a specific site and mechanism of action (Fig. **4**). This venom is antagonistic to acetylcholine and blocks neuromuscular junctions. The complete venom has a complex assortment of enzymes including phospholipases, proteinases, and other enzymes that can liquefy tissues and destroy (hemolyze) red blood cells. This causes tissue necrosis which can proceed to horrendous consequences for extremities [6, 7]. It is generally said that the mortality rate from snake bites (including rattlesnake bites) is low and this is because the amount of venom injected is low or in some cases essentially none. However, even if there are no systemic effects, the local effects of a rattlesnake bite can be severe. There usually is local swelling, local pain, and failure of the blood to clot. Muscle tissue can be damaged.

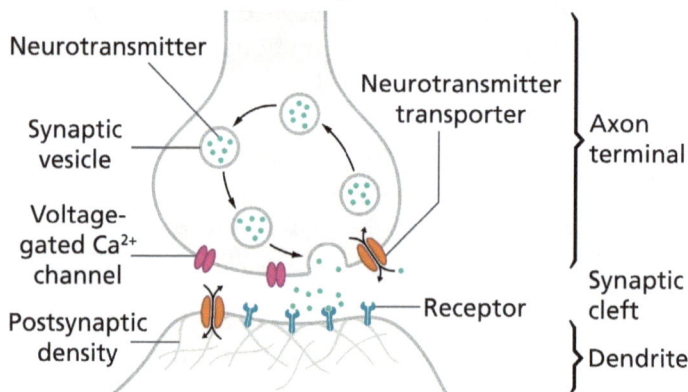

Fig. (4). Generalized view of myoneural junction, the site of action of the neurotoxic component of Mojave rattlesnake venom. Nerve impulse transmission from a nerve (top) passes the synaptic cleft to a muscle dendrite (bottom). The chemical transmitter (green dot) is released from vesicles and moves across the synapse to affect receptors and subsequently elicit muscle contraction. Specific types of pores (orange, pink) are shown and can be open or closed. Mojave rattlesnake venom has a toxin that blocks this nerve transmission.

More specifically, research has elucidated the mechanism of action of neurotoxins of the types that are found in certain rattlesnake venoms, including the Mojave rattlesnake. Clinically, presynaptic-acting neurotoxic phospholipases are components of snake venom that are said to contribute to most deaths due to snake envenomation [8]. The neurotrophic presynaptic neurotoxic activities of phospholipases of *Crotalid* venom have been correlated with specific chemical components in Mojave toxin. This venom causes complex pathological effects that include the ability to block transmission of nerve impulses across neuromuscular junctions and thus they cause paralysis. They also induce damage to muscle tissue and these mechanisms also are key factors in the pathology of rattlesnake bites [8].

Less research has been done on the neurotoxin from the Mojave rattlesnake than has been done on the neurotoxins of other pit vipers. A report in 2012 describes the variability in venoms and severity of outcomes from bites of the Mojave rattlesnake (*Crotalus scutulatus scutulatus*) [9]. The toxicity of venoms from snakes collected from Arizona and New Mexico were studied. Venoms contained toxic metaloproteinases of two types and myotoxin-A. Crotoxin is the main toxin in the venom of the South American rattlesnake (*Crotalus durissus terrificus*) and it was the first snake venom protein to be purified and crystallized [10]. The venom was found to contain an enzyme, phosphorylase A (2), and a protein without enzyme activity that was non-toxic. The toxin had the classic biological activities of the crotoxins: neurotoxicity, myotoxicity, nephrotoxicity, and cardiotoxicity. This article reviewed research that reported that *Crotalus* venom also has immune-modulatory, anti-inflammatory, antimicrobial, antitumor and analgesic activities which suggest great potential for the discovery of novel pharmacological agents. It has been reported that some Mojave rattlesnakes have 'Mojave' toxin and some do not [11] as shown by analysis of genomic differences A paper was published in 1993 that described a model for the interaction of phospholipase A2 neurotoxin from the Mojave rattlesnake with presynaptic membranes [12]. The toxin specifically impairs the release of acetylcholine at neuromuscular junctions and this occurs predominantly at the presynaptic site. The complex activity of the major toxin of the South American rattlesnake has been investigated [13]. This toxic protein has two distinct subunits: a basic phospholipase A_2 (also known as PLA_2) which has high enzyme activity. It also has an acidic non-toxic fraction and a non-enzyme component. As early as 1977, research was published describing the South American (Brazilian) rattlesnake venom and reporting its physiological effect of respiratory paralysis and peripheral neuromuscular blockade, which is unusual for a rattlesnake venom [13] [14]. These researchers determined that the effects of this venom were very similar to those of β-bungarotoxin and most likely due to the presynaptic inhibition of the mechanism responsible for the release of neurotransmitter in the nerve terminal. This highly toxic protein, crotoxin, with neurotoxic and enzymatic activity explains the reason why the Mojave rattlesnake bite is more highly venomous than most rattlesnake venoms. An interesting comparison of several snake venoms has been published [15]. This account includes the eastern diamondback rattlesnake (*Crotalus adamanteus*) and describes its relative toxicity as < 0.1 when compared to Indian cobra (*Naja naja*) as 1.0. On this scale the King cobra (*Ophiophagus Hannah*) rated 0.3. The inland taipan (*Oxyuranus microlepidotus*) ranked highest of all snakes with a relative toxicity of 50.0 (see Chapter 1).

THE COPPERHEAD SNAKE

The copperhead is native to a wide area in United States from southern New England to northern Florida and west to southern Indiana, western and southern Illinois, Missouri, southern Nebraska and southwest through much of Oklahoma and Texas [16]. More details about the range of the copperhead has been described [17]. Copperheads are said to be: "… some of the more commonly seen North American snakes. They're also the most likely to bite, although their venom is relatively mild, and bites are rarely fatal to humans" [18]. The copperhead snake (*Agkistrodon contortrix*) is a pit viper (it contains heat-sensing organs in its head) (Fig. **5**). Pit vipers also include the cottonmouth snake and the rattlesnake in North America. These snakes have elliptically-shaped (cat-like) pupils [16], a characteristic of venomous North American snakes. North American has five varieties of copperheads that differ in color and geographic distribution – some have a pinkish-tan color with distinct, dark brown markings along their back that are bow-tie shaped (hourglass shaped). The hourglass shaped markings are said to be unique to copperheads [18]. These snakes take their name from the characteristic copper-colored heads, a fact attributed by the Pennsylvania State University biology department in a report by LiveScience [18]. The coloration of copperheads allow them to blend in with leaf litter [17]. Copperheads average two to three feet in length but can grow to 43 inches [16]. Copperheads are "… semi-social snakes. While they usually hunt alone, they usually hibernate in communal dens and often return to the same den every year" [18]. An odd fact: "When touched, copperheads sometimes emit a musk that smells like cucumbers" [18]. All pit vipers are predators that coil and wait for prey to approach. Copperheads "… bite a prey animal, inject venom, then quickly release the prey. The mouse or other prey dies in minutes" [16]. This site also states that: "In Missouri, no person has died as a result of a copperhead bite. In an average year, venomous snakes bite approximately 200 people in this state, with the majority involving copperheads. In over 25 years there are no records of a person dying from the bite of any venomous snake species native to Missouri" [16]. Copperhead bites usually produce immediate and intense pain which may be followed by tingling or throbbing with nausea and signs of swelling in extremities if bites occur there. Bites seldom require antivenom [16]. All North American copperheads give birth to live young (some snakes actually lay eggs). Copperheads are often found around water and their diet is known to include lizards such as skinks, frogs, and fence lizards. It is said that "… over 90% of an adult copperhead's diet consists of mice, especially deer mice and voles. Copperheads also eat other rodents, such as house mice and young chipmunks" [16]. Young copperheads have a strange anatomy and behavior that helps them catch prey. "The last inch or so of their tails are greenish yellow with 8 or 10 small, white markings edged in black. Their colorful tails have an important use: they help them capture prey. If a small lizard or frog ventures within a few feet of a coiled baby copperhead, the copperhead will move its tail near the center of its coil, elevate it slightly and begin to wiggle or undulate the tail. To a hungry lizard that moving tail tip may look like a green caterpillar" [16].

Fig. (5). Southern copperhead.

Copperhead Venom

The venom of the copperhead snake is complex. Like all pit vipers (including the rattlesnake and the fer-de-lance) copperheads have long, hinged, hollow fangs. When they strike prey or a human victim, by voluntary action, the snake can inject toxin and they control the amount. Vipers, including the copperhead, have venoms containing a hemotoxin that affects blood. In particular, it impairs coagulation, and has necrotizing agents that cause death of tissue [19]. There are approximately 20 known different types of toxic enzymes found in snake venoms and the copperhead possesses perhaps a half-dozen of these. Each has a special function, and most are involved in tissue destruction that contributes to necrolysis. Specifically, copperheads have a phospholipase that damages muscle and nerve tissue by hydrolyzing certain types of lipids found in membranes. Copperhead venom also contains a type of chemical called a biogenic amine that disrupts normal transmission of nerve impulses and also affects signaling (communication) between certain types of cells [19]. The venom also contains a chemical with the ability to break down blood clots and thus it can cause extensive hemorrhage. The venom toxin belongs to the agents called "metallo-endopeptidases" This venom toxin contains zinc, and it is an enzyme that attacks fibrin in blood clots [20]. Many years ago, this agent was recognized as having medical uses and a therapeutic called 'alfimeprase' was produced. Alfimeparase was made in the laboratory using genetic information from the copperhead snake (*Agkistrodon contortrix contortrix*) and recombinant DNA technology to produce this protein in pure form in the laboratory. Alfimeprase is made artificially to do what a component of the snake venom does naturally – it breaks down blood clots [20]. The development of this as a drug has had a long history of failure in phase III trials, but there is continuing interest for a recombinant, truncated version of this fibrinolase for use following arterial occlusions.

THE COTTONMOUTH MOCCASIN

The cotton mouth moccasin (*Agkistrodon piscivorus*) is the only venomous water snake in North America [21]. It is distinctive in appearance with a triangular head, a thick body, and a white color inside the mouth that is prominently shown when the snake is aroused (Fig. **6**). They are pit vipers. Some local names are; black moccasin, mangrove rattler, swamp lion and water mamba [21]. They grow to a large size, up to 2 to 4 feet in length. Their color ranges from dark brown to olive, banded brown or yellow with pale bellies [21]. This site also states that juveniles have bright yellow tips on their tails that they can use as a lure to draw prey such as frogs into striking range. This site reports their range, according to the Smithsonian National Zoological Park, as the: "… Southeastern United states, from southern Virginia to Florida to eastern Texas. Another site reports that at least one of three species (eastern, western, and Florida) are found in seventeen different States" [22].

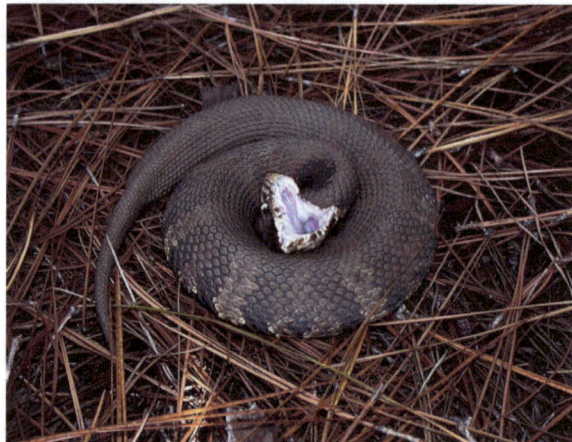

Fig. (6). Cottonmouth moccasin.

Water moccasins can be found swimming in swamps, marshes, drainage ditches, and at the edges of ponds, lakes and streams. Cottonmouth snakes are strong swimmers and the Latin name *piscivorus* refers to fish which is a favorite diet item [22]. This site provides references to much of the data it presents. Diets for large water moccasins includes lizards, salamanders, turtles, birds, other snakes, frogs and mammals [22]. Like other venomous species cottonmouth snakes obtain food by ambushing prey which they actively pursue. They also commonly eat dead animals they find. If they are successful in biting a live animal they release their bite and track it down when it is subdued by the venom [22]. On land they are found near water and fields. "They often sun themselves on branches, stones and logs near the water's edge" [21].

Cotton mouth moccasin snakes inhabit the same areas as do non-venomous water snakes [22]. This site says that these snakes mate annually and the female gives birth to up to 16 babies. In states with somewhat cold winters, cottonmouth snakes can hibernate for several months and this is sometimes done in wooded hillsides or even inside hollow stumps.

Venom of the Cottonmouth Moccasin

Cotton mouth moccasins have potent venom. It is primarily hemotoxins that destroy blood cells and components that prevent blood from clotting which leads to "hemorrhaging throughout the circulatory system wherever the venom has spread" [21]. When a cottonmouth snake bites a person, envenomation can cause temporary or even permanent damage to tissues including muscles. If an extremity is bitten, tissue necrosis may cause loss of the limb or a portion of the limb. "Anyone suffering from a water-moccasin bite should seek medical attention immediately. Antivenin is available" [21].

The venom of the cottonmouth moccasin contains several proteins including a type called lectins [22]. These lectins are calcium-dependent lectins and several types are found in a variety of venoms. These lectins can induce or inhibit platelet aggregation, and they can inhibit blood coagulation (which can occur throughout tissues where venom has spread) [22]. A lectin was purified from the venom of the Eastern cottonmouth moccasin and had a molecular weight of 16,000 Da; it was calcium-dependent and the complete amino acid sequence was determined [23]. In another report [23], three lectins were reported from venoms of three snakes - the western cottonmouth, the western diamondback, and the southern copperhead. The three lectins were indistinguishable by the test means available with a molecular weight of 28,000 Da as a dimer. In an older report (from 1982), venom from the cottonmouth moccasin had fibrinolytic activity with a molecular weight of 34,000 Da. It lacked plasminogen-activating activity [23].

The prevalence of poisonous snake bite victims at a regional trauma center in Southeastern Georgia over the 10-year period from January 1984 to January 1994 was reported [24]. This report describes: "the type of snake, grade of enve-nomation, treatment administered, morbidity and mortality, and outcome". Data was obtained over a 24-county region. Results included: "Sixty-three (63) bites in 62 victims of venomous snakebites were treated. The snake distribution was rattlesnake: 19 (30%), copperhead: 18 (29%); cottonmouth moccasins: 8 (12%), unknown: 18 (29%). Envenomation grades were Grade I: 20 (32%), Grade II: 24 (38%), Grade III: 10 (16%), and Grade IV: 9 (14%). Fourteen of 19 (74%) Grades III and IV envenomations were from rattlesnakes." Antivenin was administered in all Grade IV and half of the Grade III envenomations. Antivenin was administered within 3 hours in all but one case. Five patients had surgery. Two patients (both Grade I) developed anaphylaxes from antivenin. All patients recovered. An average of 6 snakebites were treated each year.

MAMBAS

Mambas come in several types which will be briefly described. I choose to focus on the black mamba for reasons which will become obvious. The black mamba (*Dendroaspis polylepis*) is generally acknowledged to be one of the world's deadliest snakes (Fig.7). It is said to be the fastest land snake in the world, the longest species of venomous snake in Africa, and the second longest in the world [25]. This site quotes Sara Viernum, a herpetologist: "Black Mambas are extremely toxic and very fast snakes [they are highly aggressive when threatened] ... known to strike repeatedly and [to] inject a large volume of venom with each strike." This site also reports: "Their venom is potentially lethal, and though antivenin exists, it is not widely available in the black mamba's native habitat of southern and eastern Africa... they are considered a top killer in a land where nearly 20,000 people die from snake bites every year, according to PBS's Nature" [25]. I am unable to independently verify these data.

The black mamba is one of four species of mambas.... Others are Jameson's mamba (*Dendroaspis jamesoni*), eastern green mamba (*Dendroaspis angusticeps),* and West African green mamba (*Dendroaspis viridis).* They are in the same family as coral snakes and cobras" [25]. There are also other mamba species that are smaller and somewhat less venomous. They are brilliant green and are arboreal, spending most of their life in trees. They are said to be solitary snakes and are known to drop from tree branches onto prey [25]. This sounds frightening either to experience or to observe. The black mamba actually is brownish in color with tones varying from olive to grayish, but it is named for its mouth which is "a deep, inky black" [25]. They grow to be 14 feet long (average length is approximately 8 feet. They live throughout sub-Saharan Africa, ranging from South and East Africa's savannas to rocky hills and open woodlands (reference is given), and prefer low, open spaces and are found sleeping in hollow trees, rock crevices and in empty termite mounds [25]. This site states that this snake "... can move faster than most people can run... up to 12 mph.... Over longer distances, they average about 7 mph" [25]. Providing a reference, this site states: "When threatened with no perceived available escape, these snakes will raise their upper bodies off the ground to stand erect… The front third of their bodies can rise 3 to 4 feet off the ground. Then they will spread their neck flaps [and] gape their mouths to expose the black lining … they will strike repeatedly, potentially deliver large doses of venom with each strike, and hiss loudly."

The diet of the black mamba includes small mammals and birds and they hunt by biting their prey, injecting venom, and releasing the prey. They follow the prey until it becomes paralyzed or dies, and consume it whole [25].

Fig. (7). Black mamba.

Black Mamba Venom

Quoting from the LiveScience site [25]: "Just two drops of potent black mamba venom can kill a human" (a source is cited). The venom contains neurotoxins and "… without antivenom, the fatality rate from a black mamba bite is 100 percent… Fatalities… have been documented … within as little as 20 minutes after injection… However, most known fatalities have occurred within 30 minutes to 3 hours or longer."

A protein toxin from black mamba venom has been exquisitely characterized [26]. This toxin binds with high affinity to muscular and neuronal nicotinic acetylcholine receptors and inhibits acetylcholine from binding to the receptor. This impairs neuromuscular and neuronal transmission and causes paralysis. These authors (publishing in 2014) characterized this toxin as a: acetylcholine receptor inhibiting toxin, an ion channel impairing toxin, a neurotoxin, and a postsynaptic neurotoxin. The specific name for the toxin is Alpha-elapitoxi--Dpp2d. Structurally, the protein is described as a monomer belonging to the classical, snake 3-finger toxin family, long-chain subfamily, type II alpha-neurotoxin sub-family. It contains a free carboxyl terminus. The dimeric structure of the toxin molecule is stabilized by extensive intermolecular hydrogen bonds and electrostatic interactions. However, the toxin "is predominantly monomeric under physiological conditions". It has been completely sequenced. Its molecular mass is 7,985.33±0.40 Da. The toxin is described to have the molecular function of: "Interacting selectively with one or more biological molecules in another organism (the "target" organism), initiating pathogenesis (leading to an abnormal, generally detrimental state) in the target organism... Examples include the activity of botulinum toxin, and snake venom." The authors predict that the toxin will be a valuable tool for studying the structure and function of acetylcholine receptor sites in myoneural junctions [26]. This publication illustrates the extreme sophistication that has been reached in characterizing protein toxins. This research is primarily driven by the reasonable probability that knowledge gained at the molecular level will provide increased understanding of nerve transmission including pain-blocking pharmaceuticals.

Another recent paper [27] describes the binding site and inhibitory mechanism of a specific pain relieving peptide and its binding to a specific "acid-sensing channel". This peptide is specifically described as follows: "Mambalgin-2 is a snake venom peptide that blocks acid-sensing in channels (ASICs) to relieve pain... [it] interacts with at least three different regions of ASICs and exerts both stimulatory and inhibitory effects... [it] traps the channels in the closed confirmation." Thus, it is capable of both causing and relieving pain and may allow development of blockers of ASICs with therapeutic value. This source also states that the affected channels are proton-gated and associated with nociception, fear, depression, seizure, and neuronal degeneration which suggests roles in pain and in neurological and psychiatric disorders. The word nociception comes from the Latin word meaning to harm or hurt, and it is used to describe the sensory nervous system's response to potentially harmful stimuli.

Green Mambas

Green mambas (previously mentioned) are classified as eastern green mambas or western green mambas. The eastern variety will be described. It is the smallest of the four species of mambas with adults reaching 6 to 7 feet [28]. Their range is southward from Tanzania to eastern Zimbabwe and coastal areas of Natal. It is an arboreal snake, slender, green in color, and prefers woodlands where it spends most of its time in trees (Fig. **8**). It is described as "dangerously venomous" [28]. "It is known to enter houses and often shelters in thatched roof dwellings" [29]. This reference states that this snake feeds mostly on birds, their eggs and small mammals and describes the venom. The average venom quantity is 60 to 95 mg (dry weight). It contains neurotoxins. Several other types of toxins such as mycotoxins, anticoagulants, hemorrhagins, necrotoxins, and cardiotoxins are reported not to be present. The LD_{50} is stated to be 0.45 mg/kg for mice with venom administered intravenously [30]. This site states clinical effects: "Severe envenoming possible, potentially lethal… local pain and swelling… local necrosis uncommon but can be moderate…variable non-specific effects which may include headache, nausea, vomiting, abdominal pain, diarrhea, dizziness, collapse or convulsions… common flaccid paralysis is major clinical effect… myotoxicity not likely…coagulopathy and haemorrhages unlikely to occur… renal damage unlikely… cardiotoxicity unlikely… increased sweating, salivation." Urgent antivenom therapy is said to be the most important treatment. Simple calculations show that one snake bite, with the maximum venom in the range reported (95 mg), is sufficient to kill more than 10,000 mice.

Fig. (8). Green mamba.

The western variety differs in toxicity and venom composition. The venom is primarily composed of toxins that affect both pre-synaptic and post-synaptic sites and both are neurotoxins and cardiotoxins. The LD_{50} (mouse, intraveneous) was reported to be 0.5 mg/kg [31].

COBRAS

Cobras belong, taxonomically, to the family *Elapidae.* All are venomous. They are known for their hoods and threatening, upright posture (Fig. **9**). The word cobra is derived from the Portuguese *cobra de capello* which means "hooded snake" [32]. Cobras are found throughout Africa, the Middle East, India, Southeast Asia, and Indonesia. There is controversy about how to define species of cobras and the estimated number ranges from 28 to 270 [32]. Cobras reside in trees and are the only snake species known to build nests for their young [32]. Cobras have round pupils, smooth scales, and vary widely in color including red, yellow, black, mottled, banded, and other patterns. Most species grow to be large – some over 6 feet in length, and one, the Ashe's spitting cobra is said to reach 9 feet and the king cobra even longer (see below).

Fig. (9). King cobra.

Most cobras can raise the upper third of their body to "stand erect" in a threatening posture along with hooding and hissing. They reproduce by laying eggs which are defended [32, 33]. Classically, the mongoose is known as the chief opponent of the cobra. Curiously, the cobra is said to be deaf to ambient noise and the popular conception of this snake's association with snake charmers results because they are "enticed by the shape and movement of the flute" [32]. The king cobra is reported to have very good vision, unusual in snakes – they can identify prey at up to 300 feet away [33]. They have "one of the longest forked tongues of all snake species… they are able to smell with it."

The king cobra (*Ophiophagus Hannah*) is the only member of its genus (as determined by most scientists). The king cobra is reported to be the longest venomous snake species in the world and can reach a length of 18 feet. Since they can elevate a third of their body, they could actually stand eye-to-eye with a human [32, 33].

Snake Charmers

The ancient art of "snake charming" involves the cobra. The practice occurs most frequently in India, but also is practiced in other Asian nations [34]. By playing a flute-like instrument the snake charmer apparently "hypnotizes" the cobra which is a symbol of evil. It is said [34] that snake charming originated in ancient Egypt, although the typical exhibit of today arose in India. It has declined, mostly due to a law passed in 1972 in India assessing a seven-year prison term for owning or selling snakes. It is also said that sometimes the cobra's fangs or venom glands are removed, or the snake's mouth is sewn shut. "Hinduism has long held serpents sacred… In pictures, the cobra sits poised ready to protect many of the gods. By inference, traditionally Indians tend to consider snake charmers holy men influenced by the gods [34].

Cobra Venom

"Cobra bites can be fatal, especially if left untreated" [32]. Cobra venom contains neurotoxin. Envenomation produces vision problems, difficulty in swallowing and speaking, skeletal muscle weakness, difficulty breathing including respiratory failure (which can occur in as little as 30 minutes), vomiting, abdominal pain, necrosis developing at the site of the bite, and failure of the blood to clot [32]. The bite of a king cobra contains enough venom to kill 20 people [32]. The king cobra is said to be "… one of the most dangerous of all snakes in the world" [33]. It is known to bite repeatedly and can repeat strikes within seconds.

In 2006 a paper was published that gave an account that was said to be the first known case of king cobra envenomation in the United Kingdom [35]. The patient (a 22-year-old man) was bitten by a king cobra on his left index finger. He complained of shortness of breath and was unable to swallow within 30 minutes of arrival at the hospital. He developed hypertension (systolic blood pressure of 200-220, and peaking at 250 mm Hg). He required tracheal intubation, ventilation and treatment with 20 vials of antivenom (mono-specific for king cobra). He developed anaphylaxis with bronchospasm and hypotension that required adrenalin infusion. He was discharged 24 hours after the bite to the care of a plastic surgeon for management of isolated ischemia of the bitten finger.

This site provides a brief synopsis about venom and envenomation [35]. "In most elapid snakes the volume of venom injected is under voluntary control, leading to a variable degree of clinical effects… The main constituent of king cobra venom is a postsynaptic neurotoxin, and a single bite can deliver up to 400-500 mg of venom. The toxicity of venom… is 1.91 mg/kg [mouse LD_{50}]… so that one bite may contain up to 15,000 LD_{50} mice doses. In comparison, the world's most venomous snake, the Australian elapid small-scaled snake (*Oxyuranus microlepidotus*), can deliver up to 100 mg of venom with an LD_{50} for mice of 0.01 mg/kg, giving up to 500,000 LD_{50} mice doses per bite." This site provides a reference for the toxicity data [36].

Brown Snakes

In this chapter, the brown snake refers to the World's second most venomous snake, not the brown snake of North America – a shy, non-venomous creature. The venomous brown snake's range is Australia, Papua New Guinea and West

Papua [37, 38]. It includes nine species of the genus *Pseudonaja*. All are venomous; they are long and slender with narrow heads; typically they are uniformly brown but some are patterned (Fig. **10**). Some have black heads. They are typically 3 to 6 feet but can reach 8 feet long [37]. The eastern brown snake is said to be responsible for most human bites by the brown snake in Australia. It thrives in a variety of habitats, particularly grasslands, woodlands, and pastures, and it even invades urban areas – thus, it is responsible for the majority of recorded snakebites on Australia (approximately 5 per year from all brown snake species) [37]. Brown snakes mostly eat small mammals, frogs, small birds, and reptiles such as skinks and geckoes [37, 38]. They use both venom and constriction to kill their prey. Males engage in ritualistic combat over females and braid their bodies together with attempts to push the opponents head to the ground. Females lay eggs and the mother snake may stay with the eggs for up to five weeks [37].

Fig. (10). Pseudechis australis (brown snake).

Brown Snake Venom

It is said that because this snake thrives in areas populated by humans, it is encountered frequently by humans [38]. This snake often reacts defensively when surprised and will strike without hesitation. Relative to other elapids, the fangs of Eastern Brown snakes are small (only about 3 mm) and the average venom yield is about 4 mg (although a yield of 67 mg was recorded) [38]. The venom is very potent and contains toxins that are active as a presynaptic neurotoxin, procoagulant, cardiotoxin and nephrotoxin. Envenomation leads to progressive paralysis and uncontrolled bleeding into the brain.

It is reported [37] that "According to most standards, including the Commonwealth Serum Laboratory Lethal Dose count, the Eastern brown snake is the second most toxic snake in the world (The inland Taipan snake, also found in Australia, is No. 1). The Western brown snake is the 10th most toxic snake." The venom of Eastern brown snakes contains primarily neurotoxins, specifically textilotoxin which "… has the highest lethality of any known snake venom neurotoxin. Their venom also contains strong coagulants as well as cardiotoxins and nephrotoxins. A bite from an Eastern brown snake can result in dizziness, convulsions, renal failure, cardiac arrest, paralysis, and uncontrolled bleeding" [37]. Because the fangs are small and the bite may be almost painless initially, victims need to be wary and seek medical attention quickly when bitten by this snake. "This species has the unfortunate distinction of causing more deaths from snake bite than any other species of snake in Australia" – this from the Australian Museum site [38].

A component has been isolated and studied from the Australian brown snake venom that has potential therapeutic activity as a systemic anti-bleeding agent. Textilinin-1 is one such experimental drug to decrease blood loss during surgery, and it is a possible replacement for existing drugs that have significant side-effects. [39]. "The more rapidly reversible inhibition of plasmin by textilinin-1 in comparison with aprotinin [an existing drug] would be expected to result in a faster recovery of plasmin activity after cessation of treatment, leading to a decreased tendency to develop post-operative thrombosis" [39].

A case report of envenomation by a brown snake is informative [40]. This site states that envenomation causes consumptive coagulopathy, thrombocytopenia, microangiopathic haemolytic anaemia, and acute renal failure. Consumption coagulopathy is a common complication of envenomation by the brown snake. A study in 2006 reported that one vial of antivenom was found to be sufficient to bind and neutralize all venom in patients with severe brown snake envenomation [41].

Venom from the Eastern brown snake is stated "to be the second most toxic snake venom known and it is the most common cause of death from snake bite in Australia. This venom (Fig. **11**) is known to contain a prothrombin activator complex, serine proteinases inhibitors, various phospholipase A2s, and pre-and post-synaptic neurotoxins" [42]. These researchers reported that the venom contained "an abundance of glycoproteins … multiple isoforms of mammalian coagulation factors that comprise a significant proportion of the venom".

A component, named pseutarin C (a prothrombin activator) was isolated from Australian brown snake venom [43]. This toxin converted prothrombin to thrombin by cleaving two peptide bonds with action very similar to the mammalian factor which performs this function. The authors state: "… pseutarin C is the first venom procoagulant protein that is structurally and functionally similar to [the mammalian blood factor that performs this function naturally]". A clear, scientific understanding is not available to explain the presence of these specifically-acting biological agents in this snake.

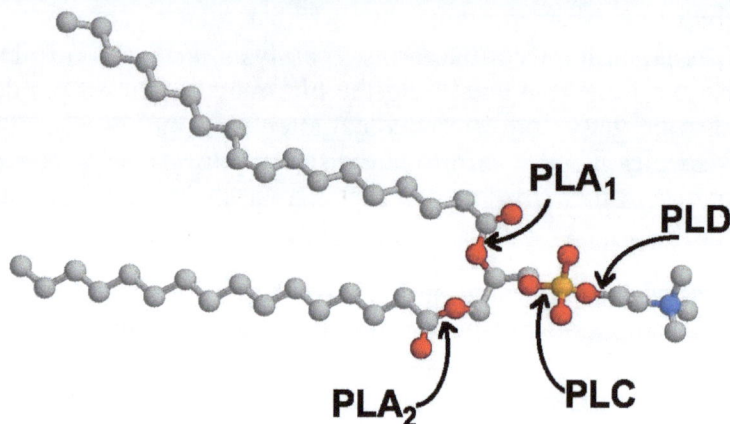

Fig. (11). Molecular sites of action of phosphorylases. PLA_2 is from elapid snake venom and cleaves a particular site in phospholipid, as shown.

Inland Taipan

The toxic venom of the inland taipan was addressed in Chapter 1. The biology of this snake is described here including a bit more about the venom. The inland Taipan, *Oxyuranus microlepidotus* (Fig. **12**), is most often stated to be "the world's most venomous snake." This is difficult to prove scientifically because there is ambiguity in this definition, many laboratory techniques are involved, and there is variability in individual snakes. The website "Australian Museum" [44] states: "… the Inland Taipan is far from the most dangerous." They explain and defend the statement by saying: Unlike its congener, the common and fiery-tempered Coastal Taipan, this shy serpent is relatively placid and rarely encountered in its remote, semi-arid homeland." Significantly, another site [45] says "No human fatalities are recorded, but certainly the inland taipan is large enough to have a fatal bite like that of its close relative the coastal taipan, *Oxyuranus scutellatus.*" It is like the old saw "let us first define our terms, then we will argue." With enough data, done carefully enough, it is possible to determine which venom component is the most lethal – that is if we define the test victim species. It is possible to compare between species, from a mouse to a human for example, but as we do so, the reliability of conclusions becomes lower. The effect on humans is often the focus of interest regarding venomous snakes. In toxicology, this is similar to the distinction made between a hazard and a risk. A hazardous chemical might be very poisonous, for example. Humans are not at risk, however, if there is no access to the chemical. Another analogy may serve. A rancher may have a poisonous weed on his land. If it is fenced-off and his cows have no access to those weeds, they pose no risk to the herd. A large, aggressive snake with a very powerful venom stored in a large quantity, with huge fangs to deliver it, and a fierce temperament to bite, may rarely kill humans if the snake keeps itself remote from humans. There are a large number of variations of the described variables. Thus it is obvious that ranking venoms has as many answers as do rankings of football teams at play-off season. However, I do engage by giving the usual assessments, and in Chapter 1, I propose ranking venom components and pure chemicals on the basis of the number of molecules required to kill a mouse – and with this equated to the human. This ranking, as explained more fully in Chapter 1, is different from the standard ranking based on mass (not number of molecules).

Fig. (12). Inland Taipan commonly said to be the world's most venomous snake.

The inland taipan is sometimes called the fierce snake, small-scaled snake, or lignum snake [44]. It is described as: "A medium to large snake with a robust build and a deep, rectangular-shaped head." The back of the snake varies in color from "… pale fawn to yellowish-brown to dark brown with the head and neck being … darker than the body… Eyes are large, with a very dark iris and round pupils" [44]. Its average size in approximately 6 feet, with the largest reliably measured and reported in the literature was 250 cm (a little more than 8 feet) [44]. It further states that it can be confused with the eastern and western brown snake, and that the inland taipan occurs in the "channel country of south-western Queensland and north-eastern South Australia (with less reliable sightings elsewhere). Curiously "…though it was described in 1879, it was not rediscovered until 1967 when its bite almost killed the first person to see it in almost a century" [45]. It is said to favor deep clays and loams of floodplains but the snake is also found in dunes and rocky outcrops; these snakes shelter in soil cracks and crevices and in holes and animal burrows [44]. [They] "feed entirely on small to medium-sized mammals, particularly the long-haired rat … as well as the introduced House Mouse… and various small dasyurids" (Australian marsupials) [44]. This site says that "…venom acts so quickly that the snake can afford to hold on to its prey instead of releasing (to avoid injury) and waiting for it to die." This appears to be a significant feature of this snake, and apparently is the result of the extremely quick-acting and lethal venom. Eggs are laid by females and require 9 to 11 weeks to hatch.

NOTES

[1] Ralph Waldo Emerson [1803-1882] Journal, 11 April 1834.

REFERENCES

[1]   rattlesnake | snake | Britannica.com [Internet]. [cited 2017 Nov 26]. Available from: https://www.britannica.com/ animal/rattlesnake

[2]   Rattlesnake | San Diego Zoo Animals & Plants [Internet]. [cited 2017 Nov 26]. Available from: http://animals.sandiegozoo.org/animals/rattlesnake

[3]   Centennial by James A. Michener Study Guide | Arapaho | Magma [Internet]. [cited 2017 Nov 26]. Available from: https://www.scribd.com/document/157923999/Centennial-by-James-A- Michener-Study-Guide

[4]   True Grit | NEA [Internet]. [cited 2017 Nov 26]. Available from: https://www.arts.gov/ partnerships/nea-big-read/true-grit

[5]   The Rattler. Literary Analysis. [Internet]. [cited 2017 Nov 26]. Available from: http://www.markedbyteachers.com/gcse/english/the-rattler-literary-analysis-the-rattler-is-a-compelling -chronicle-of-a-man-killing-a-rattlesnake-although-this-man-narrates-the-story-i- -is-told-from-the-perspective-of-the-snake.html

[6]   Rattlesnake Bites and Symptoms - DesertUSA [Internet]. [cited 2017 Nov 26]. Available from: https://www.desertusa.com/reptiles/rattlesnake-bites.html

[7]   Mechanisms of Venom Toxicity [Internet]. [cited 2017 Nov 26]. Available from: http://jrscience.wcp.muohio.edu/studentresearch/costarica03/venom/mechanisms.htm

[8]   Gasanov SE, Dagda RK, Rael ED. Snake Venom Cytotoxins, Phospholipase A2s, and Zn(2+)-dependent Metalloproteinases: Mechanisms of Action and Pharmacological Relevance. J Clin Toxicol [Internet] NIH Public Access , 2014 Jan 25; [cited 2017 Nov 26];4(1): 1000181. Available from: http://www.ncbi.nlm.nih.gov/pubmed/24949227

[9]   Massey D, Calvete J, Sanchez E, Richards K, Curtis RBK. Venom variability and envenoming severity outcomes of the Crotalus scutulatus scutulatus (Mojave rattlesnake) from Southern Arizona. J Proteomics [Internet] Elsevier , 2012 May 17; [cited 2017 Nov 27];75(9): 2576-87. Available from: https://www.sciencedirect.com/science/article/pii/S1874391912001340

[10]   Sampaio SC, Hyslop S, Fontes MRM, Prado-Franceschi J, Zambelli VO, Magro AJ, *et al*. Crotoxin: Novel activities for a classic β-neurotoxin. Toxicon [Internet] , 2010 Jun; [cited 2017 Nov 27];55(6): 1045-60. Available from: http://linkinghub.elsevier.com/retrieve/pii/S0041010110000127

[11]   Woodbridge B, Pineda G, Banuelas-ornelas J, Dagda R, Gasanov S, Rael E, *et al*. Mojave rattlesnakes (Crotalus scutulatus scutulatus) lacking the acidic subunit DNA sequence lack Mojave toxin in their venom. Comp Biochem Physiol Part B Biochem Mol Biol [Internet] Pergamon , 2001 Sep 1; [cited 2017 Nov 27];130(2): 169-79. Available from: https://www.sciencedirect.com/science/article/pii/S109 6495901004225

[12]   Délot E, Bon C. Model for the interaction of crotoxin, a phospholipase A2 neurotoxin, with presynaptic membranes. Biochemistry [Internet] , 1993 Oct 12; [cited 2017 Nov 27];32(40): 10708-3. Available from: http://www.ncbi.nlm.nih.gov/pubmed/8399216

[13]   Pereanez J, Gomez I, Patino A. Relationship between the structure and the enzymatic activity of crotoxin complex and its phospholipase A2 subunit: An in silico approach. J Mol Graph Model [Internet] Elsevier , 2012 May 1; [cited 2017 Nov 27];35: 36-42. Available from: http://www.sciencedirect.com/science/article/pii/S1093326312000071

[14]   Chang CC, Lee JD. Crotoxin, the neurotoxin of South American rattlesnake venom, is a presynaptic toxin acting like? -bungarotoxin. Naunyn Schmiedebergs Arch Pharmacol [Internet] Springer-Verlag , 1977 Jan; [cited 2017 Nov 27];296(2): 159-68. Available from: http://link.springer.com/10.1007/BF00508469

[15]   Relative Toxicity [Internet]. [cited 2017 Nov 27]. Available from: http://venomsupplies.com/toxicity/

[16]   Missouri Copperheads | Missouri Department of Conservation [Internet]. [cited 2017 Nov 27]. Available from: https://mdc.mo.gov/conmag/1999/05/missouri-copperheads

[17]   Copperhead Snake.com [Internet]. [cited 2017 Nov 27]. Available from: http://www.copperhead-snake.com/

[18]   Copperhead Snakes: Facts, Bites; Babies [Internet]. [cited 2017 Nov 27]. Available from: https://www.livescience.com/ 43641-copperhead-snake.html

[19]   Types of snake venom [Internet]. [cited 2017 Nov 27]. Available from: http://www.chm.bris.ac.uk/webprojects2003/stoneley/types.htm

[20]   Alfimeprase. Drugs R D [Internet] , 2008 [cited 2017 Nov 27];9(3): 185-90. Available from: http://www.ncbi.nlm.nih.gov/ pubmed/18457471

[21]   Facts About Water Moccasin (Cottonmouth) Snakes [Internet]. [cited 2017 Nov 27]. Available from: https://www.livescience.com/43597-facts-about-water-moccasin-cottonmouth-snakes.html

[22]   Cottonmouth Snake - facts [Internet]. [cited 2017 Nov 27]. Available from: http://www.cottonmouthsnake.net/

[23]   Komori Y, Nikai T, Tohkai T, Sugihara H. Primary structure and biological activity of snake venom lectin (APL) from Agkistrodon p. piscivorus (Eastern cottonmouth). Toxicon [Internet] Pergamon , 1999 Jul 1; [cited 2017 Nov 27];37(7): 1053-64. Available from: https://www.sciencedirect.com/science/article/pii/S0041010198002396

[24]   Rudolph R, Williams J, Neal G, McMahan A. Snakebite treatment at a Southeastern regional referral center [Internet]. [cited 2017 Nov 27]. Available from: https://www.researchgate.net/publication/15536630_Snakebite_treatment_at_a_Southeastern_regional_referral_center

[25]   Szalay J. Black Mamba Facts [Internet]. [cited 2017 Nov 27]. Available from: https://www.livescience.com/43559-black-mamba.html

[26]   Wang C-IA, Reeks T, Vetter I, Vergara I, Kovtun O, Lewis RJ, *et al.* Isolation and Structural and Pharmacological Characterization of α-Elapitoxin-Dpp2d, an Amidated Three Finger Toxin from Black Mamba Venom. Biochemistry [Internet] , 2014 Jun 17; [cited 2017 Nov 28];53(23): 3758-66. Available from: http://pubs.acs.org/doi/10.1021/bi5004475

[27]   Salinas M, Besson T, Delettre Q, Diochot S, Boulakirba S, Douguet D, *et al.* Binding site and inhibitory mechanism of the mambalgin-2 pain-relieving peptide on acid-sensing ion channel 1a. J Biol Chem [Internet] American Society for Biochemistry and Molecular Biology , 2014 May 9; [cited 2017 Nov 28];289(19): 13363-73. http://www.ncbi.nlm.nih.gov/pubmed/24695733

[28]   Green Mamba [Internet]. [cited 2017 Nov 28]. Available from: http://www.reptilesmagazine.com/Snake-Species/Green-Mamba/

[29]   WCH Clinical Toxinology Resources [Internet]. [cited 2017 Nov 28]. Available from:

http://www.toxinology.com/fusebox.cfm?fuseaction=main.snakes.display&id=SN0168

[30]   Shupe S, Ed. Venomous Snakes of the World: A Manual for Use by US Amphibious Forces. 1st ed., NY 2013.

[31]   Spawls S, Branch B. . The Dangerous Snakes of Africa. Blandford 1995; pp. 51-2.

[32]   Facts About Cobras [Internet]. [cited 2017 Nov 28]. Available from: https://www.livescience.com/43520-cobra-facts.html

[33]   King Cobra | National Geographic [Internet]. [cited 2017 Nov 28]. Available from: https://www.nationalgeographic.com/animals/reptiles/k/king-cobra/

[34]   Snake charming - New World Encyclopedia [Internet]. [cited 2017 Nov 28]. Available from: http://www.newworldencyclopedia.org/entry/Snake_charming

[35]   Veto T, Price R, Silsby JF, Carter JA. Treatment of the first known case of king cobra envenomation in the United Kingdom, complicated by severe anaphylaxis. Anaesthesia [Internet] Blackwell Publishing Ltd , 2007 Jan 1; [cited 2017 Nov 28];62(1): 75-8. Available from: http://doi.wiley.com/10.1111/j.1365-2044.2006.04866.x

[36]   Broad A, Sutherland S, Coulter A. The lethality in mice of dangerous Australian and other snake venom. Toxicon [Internet] Pergamon , 1979 Jan 1; [cited 2017 Nov 28];17(6): 661-4. Available from: http://www.sciencedirect.com/science/article/pii/0041010179902459?via%3Dihub

[37]   Facts About Brown Snakes [Internet]. [cited 2017 Nov 28]. Available from: https://www.livescience.com/53580-brown-snakes.html

[38]   Eastern Brown Snake - Australian Museum [Internet]. [cited 2017 Nov 28]. Available from: https://australianmuseum.net.au/eastern-brown-snake

[39]   Millers E-KI, Johnson LA, Birrell GW, Masci PP, Lavin MF, de Jersey J, et al. The structure of Human Microplasmin in Complex with Textilinin-1, an Aprotinin-like Inhibitor from the Australian Brown Snake. Massiah M, editor PLoS One [Internet] , 2013 Jan 15; [cited 2017 Nov 28];8(1): e54104. Available from: http://dx.plos.org/10.1371/journal.pone.0054104

[40]   Isbister GK, Little M, Cull G, McCoubrie D, Lawton P, Szabo F, et al. Thrombotic microangiopathy from Australian brown snake (Pseudonaja) envenoming. Intern Med J [Internet] Blackwell Publishing Asia , 2007 Aug 1; [cited 2017 Nov 28];37(8): 523-8. Available from: http://doi.wiley.com/10.1111/j.1445-5994.2007.01407.x

[41]   Isbister K, O'Leary M, Schneider J, Simon G, Brown B. Efficacy of antivenom against the procoagulant effect of Australian brown snake (Pseudonaja sp.) venom: *In vivo* and *in vitro* studies. Toxicon [Internet] Pergamon , 2007 Jan 1; [cited 2017 Nov 28];49(1): 57-67. Available from: https://www.sciencedirect.com/science/article/pii/S0041010106003436

[42]   Birrell GW, Earl S, Masci PP, de Jersey J, Wallis TP, Gorman JJ, et al. Molecular Diversity in Venom from the Australian Brown Snake, Pseudonaja textilis. Mol Cell Proteomics [Internet] , 2006 Feb; [cited 2017 Nov 28];5(2): 379-89. http://www.mcponline.org/lookup/doi/10.1074/ mcp.M500270-MCP200

[43]   Rao V, Kini R. Pseutarin C, a prothrombin activator from Pseudonaja textilis venom: Its structural and functional similarity to mammalian coagulation factor Xa-Va complex [Internet]. [cited 2017 Nov 28]. Available   from: https://www.researchgate.net/publication/11096114_Pseutarin_C_a_prothrombin_activator_from_Pse udonaja_textilis_venom_Its_structural_and_functional_similarity_to_mammalian_coagulation_factor_ Xa-Va_complex

[44]   Beatson C. Inland Taipan, Oxyuranus microlepidotus - Australian Museum [Internet]. [cited 2017 Nov 28]. Available from: https://australianmuseum.net.au/inland-taipan

[45]   Inland Taipan / Fierce Snake [Internet]. [cited 2017 Nov 28]. Available from: http://www.reptilesmagazine.com/Snake-Species/Inland-Taipan-/-Fierce-Snake/

CHAPTER 9

A Potpourri of Poisons

Abstract: In creating this book, most poisons of interest naturally fall into reasonably discreet groups. However, there are many other interesting poisonous substances that do not deserve a full chapter and also are disparate in nature. Selecting poisons from this group of chemicals is fraught with risks – being too inclusive or too exclusive. In the end I made choices based on several factors: my interests (I admit it); the extent to which the substance actually has poisoned people; its appearance in literature, including murder mysteries; extensive public awareness of the agent; and its potency. Medicines are a great boon to humankind and I hesitated to include them in a list of poisons. In the end, I justified their inclusion because awareness is useful knowledge and because I sense there is curiosity about this aspect of many highly-prescribed medicines. Several toxic chemicals are included because they are deserving but not a good fit elsewhere. A host of chemicals are sufficiently toxic to be included in a compendium, which this book is not. Environmental pollution raises concern for toxins in our air, water and food. These poisons are present as trace amounts and may be involved in chronic health effects but do not belong in this account. Some chemicals have, unfortunately by my thinking, been unfairly accused and even damned by popular media and lawsuits. I will simply leave these out of my list.

"The coward's weapon, poison."[1]

Keywords : Anesthetics, Bacterial Toxins, Blood Coagulation, Brodifacoum, Cancer Chemotherapeutics, Carbon Monoxide, Curare, Doxorubicin, Ethanol, Ethylene Glycol, Fentanyl, Hyperbaric Oxygen, Lethal Dose, LC_{50}, LD_{50}, Methanol, Opioid Epidemic, Painkillers, Paraquat, Pesticides, Pharmaceuticals, Rodenticides, Statins, Tubocurarine, Strychnine, Therapeutic dose, Toxic Dose, Vitamin K, Warfarin.

INTRODUCTION

This chapter contains a collection of disparate poisons, each deserving of its place in this book; however, they share most prominently only the fact that they are poisonous. The qualification criteria (beyond this limit) are arbitrary and influenced by my interests. The words of John Fletcher cited in footnote 1 summarize the focus of a topic running through this chapter (and through the book also). Another topic of this chapter is encompassed by other words attributed to Fletcher: "I find the medicine worse than the malady" (I am unable to identify the specific source). I hesitated to include prescription medicines in this list because of their value in restoring health. However, their power in so doing is often because they have a great power to destroy. The skill of the scientist and the doctor are required to their fullest extent to make medicines that are effective and safe to administer.

MEDICINES

There is a reasonable objection to including medicinals in a book about poisons. However, there is no class of chemicals more deserving of the admonition of Paracelsus, paraphrased as "the dose makes the poison" (Chapter 1). In this light, it is perhaps understandable that chemical agents capable of affecting the intricate fabric of life and the daily activities of living cells would, in excess, be fully capable of great harm. Opioids force themselves to the top of the drug list because they are currently causing an epidemic of pain throughout society, the very disorder they are designed to alleviate in the individual patient.

Pain Medications

Currently on an annual basis, tens of thousands of mostly ordinary folks and an even larger number of friends and relatives are tragically and lethally affected–either addicted or killed–by medicines designed to alleviate, not exacerbate, pain and suffering. The tragic effects are spreading in a plague-like epidemic to affect policemen and drug enforcement officials, physicians, and perhaps least of all to politicians who seem bewildered and effete in applying laws and regulations to change behavior.

Prescription opioid overdose is a significant component of this tragedy. The White House Office of National Drug Control Policy, the Department of Health and Human Services, the Department of Justice, the Department of Veterans Affairs, and the Department of Defense seem powerless in their efforts over the past 15 years to stem the opioid epidemic. The latest statistics (reported for 2015) tell an incredible story of unintended consequences: 12.5 million people misused prescription opioids, 2.1 million people misused prescription opioids for the first time, 33,091 people died from overdosing opioids, 828,000 people used heroin, 135,000 people used heroin for the first time, 12,989 deaths were attributed to overdosing on heroin, and the economic cost (2013 data) was estimated to be 78.5 billion dollars [1].

To provide a reasonable perspective, the focus of this section will be primarily on opioids as prescribed for pain and not on illicit drug trafficking or on the broad array of other pain killers. How these agents are intended to work at the tissue and biochemical levels will be the focus and we will only incidentally address the psychology of pain. Several venoms that previously were addressed can cause pain at sites and by mechanisms that are germane for understanding opioids as pharmaceuticals. It is hoped that the reader is already prepared, based on information in prior chapters, with essential aspects of nerve impulse transmission and can apply this knowledge to pain. An overview of the site and mechanism of action of opioids may be helpful (Fig. **1**).

A recent (2017) review of the impact of pain on overall survival in cancer patients points to the necessity of pain medication in this complex disease [2]. This review concluded (in part): "Pain may be associated with shorter survival time in patients with cancer, but the mechanism for this relationship is unknown. The available evidence is insufficient to definitively determine if pain independently influences survival in patients with breast, colorectal, or lung cancer." Regardless of this, pain control is essential in treating cancer and less addictive and safer drugs are needed. The mechanism of opiate addiction is different from other drugs [3]. Chronic morphine exposure has an opposite effect on the brain, compared to cocaine when studied in mice and this study reported that a protein in the brain (called brain-derived neurotrophic factor) was increased in cocaine addiction but inhibited in opioid addiction. The report also stated: "Morphine creates reward by inhibiting [the neurotrophic factor], whereas cocaine acts by enhancing [the neurotrophic factor] activity." This points to the complexity at the biochemical level of the action of these drugs. The lethal effects primarily result from severe depression of respiration or cardiac arrest.

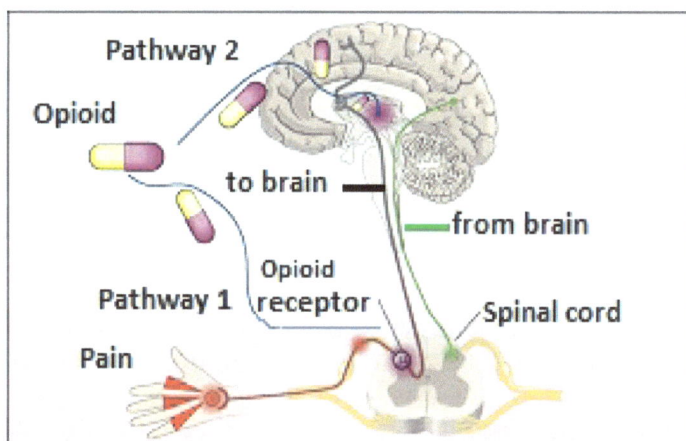

Fig. (1). Overview of the mechanism of pain and the action of opioids to alleviate it; see text for details.

Opioids are a class of drugs that includes the illegal drug heroin, synthetic opioids such as fentanyl (also known as fentanil), and the prescription drugs for pain relief such as oxycodone, hydrocodone, codeine, morphine, and others. These drugs are chemically related and interact with opioid receptors on nerve cells in the body and in the brain. When taken as appropriately prescribed for short periods of time they are generally regarded as safe. However, they produce euphoria along with pain relief and can be misused [4]. This can lead to dependence, which is beyond the scope of the subject of this book.

The toxicities of opioids are relevant. Fentanyl is a narcotic analgesic more potent than morphine [5]. It is an average sized molecule with a molecular weight of 336.471 grams per mol. It acts predominantly at what are called μ-opiate receptors. In addition to analgesia (pain relief), fentanyl produces drowsiness and euphoria. Common side effects include nausea, dizziness, vomiting, fatigue, headache, constipation, anemia, and peripheral edema. Overdoses cause respiratory depression which is reversed by the antidote naloxone. Sudden death can occur because of cardiac arrest or severe anaphylactic reaction [5]. The estimated lethal dose of fentanyl in humans is 2 mg. The recommended serum concentration for analgesia is 1-2 ng/ml and for anesthesia it is 10-20 ng/ml. "Blood concentrations of approximately 7 ng/ml or greater have been associated with fatalities where poly-substance use was involved" [5]. Fentanyl can be given in skin patches. Fentanyl in such patches or in used patches discarded inappropriately has been misused with fatal consequences [5]. This site also states data comparing the toxicity of fentanyl with other opioids, based on animal tests.

Fentanyl is said to be "… a narcotic analgesic with a potency at least 80 times that of morphine." This refers to its narcotic and or analgesic effects which is not necessarily the same as its lethal dose comparison. Other data from this site provides additional comparisons: "Animal tests… showed that the analgesic potency of fentanyl was 470 times that of morphine and other fentanyl derivatives were even more potent." This site further stated: "It is difficult to be certain that this increased analgesic potency means … the overdose potential of these analogues is also increased by the same margin" [5]. Opioid lethal doses have been compared [6]: "Morphine doses over 200 mg are considered to be lethal… the lethal dose of heroin is generally reported as being between 75 and 375 mg… the lethal dose for hydrocodone is generally stated to be around 90 mg… A single dose of 40 mg or more of oxycodone may produce lethal effects in some individuals… the lethal dose of fentanyl is generally stated to be 2 mg… Based on the above figures, one can calculate that *the lethal dose for fentanyl is approximately 100 times less than the lethal dose for morphine*." [Italics in original].

Another way of expressing the toxicity of drugs is to compare the active dose with the lethal dose of the drug. For example, alcohol has a lethal dose that is approximately 10-times the 'active dose'. Thus, for alcohol the ratio of the active dose to the lethal dose is 0.1 (it takes 10-times the active dose to be lethal). If we substitute blood level for dose– a bit of a stretch – and we use 0.05 BAC (blood alcohol concentration) as a proxy for active 'dose' (this amount is almost universally said to produce 'effects'), then the lethal dose would be 10-times this, or 0.5 BAC. This is usually considered to be in the upper lethal range.

For addictive drugs, the dependence increases upwards. The dependence potential is subjective and not measurable except in general terms. Addictive drugs are categorized by the Food and Drug Administration according to terminology called the Drug Schedule on a scale of I to IV which is useful for separating therapeutic drugs from drugs of abuse by categories. For example, schedule I drugs have a very high potential for addiction (and thus, for abuse) and no known therapeutic uses.

Statins

Disclaimer: no reader of this book should make any personal decision about taking statins based on anything in this book. The information about the benefits and side-effects of statins are fully available, including to all physicians who prescribe them based on many factors. I am not a physician. I have read hundreds of publications in this area including the current recommendations of authoritative medical voices. Efficacy of the statins will not be reviewed here because my purpose is simply to provide information relative to the toxicity of statins in relation to their mechanisms of action and not to review their benefits.

Statins (first known use of this word was in 1985) are defined [7]: "Any of a group of drugs (such as lovastatin and simvastatin) that inhibit the synthesis of cholesterol and promote the production of LDL-binding receptors in the liver resulting in a usually marked decrease in the level of LDL and a modest increase in the level of HDL circulating in the blood plasma." The biological mechanism of action of statins thus is to inhibit a specific enzyme required for the biosynthesis of cholesterol. This enzyme that is inhibited is called HMG-CoA reductase (Fig. **2**). Like all enzymes, its name ends in 'ase'. Inhibition of this enzyme by statins reduce the circulating level of cholesterols, including a type called low density cholesterol (LDL). LDL is associated with certain cardiovascular pathologies when circulating levels are elevated. Thus, statins act by competitively inhibiting the first enzyme in a special biosynthetic pathway in the liver. This pathway synthesizes cholesterol which is essential but associated with harmful effects when in excess, and it also synthesizes coenzyme Q10 which is essential in cells for energy production. There is evidence that statins produce other effects that are beneficial; this is beyond the scope of this book.

CHOLESTEROL, CoQ 10 BIOSYNTHESIS

Acetyl CoA

⇩

HGMA CoA

⇩   ⇐   **Statin Action Site**

Mevalonate

⇩

Farnesial pyrophosphate   ⇒   **Coenzyme Q10**

⇩

Squalene

⇩

Cholesterol

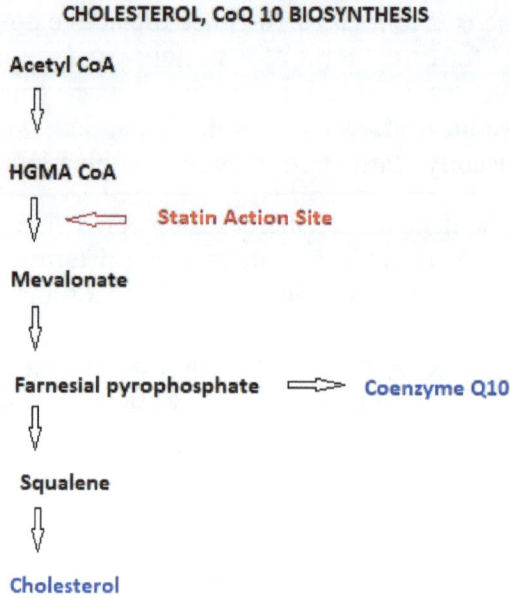

Fig. (2). Statins inhibit (poison) an enzyme required for the biosynthesis of both cholesterol and coenzyme Q10 (required for mitochondrial production of ATP, the energy source of cells).

The toxicity of statins is not well understood. There is no published LD_{50} for statins but the value surely is high compared to the therapeutic dose. Except for the possibility of an allergic reaction that could remotely lead to anaphylaxis, statins do not have significant acute toxicity. Muscle toxicity syndromes are associated with statin therapy but opinions vary regarding their frequency and severity [7]. It is relevant that statins inhibit the biosynthetic pathway required for cells to make a coenzyme called coenzyme Q10 (Fig. **2**). Coenzyme Q10 is required by every cell in the body that uses oxygen in the production of cellular energy in the form of adenosine triphosphate (ATP). Long-term treatment with simvastatin was shown to decrease maximal mitochondrial oxidative phosphorylation (the main process that generates ATP) [7]. The nearest one can come to assessing the toxicity of statin drugs (no relevant LD_{50} toxicity data are in the published literature, to my knowledge) is to examine lethal rhabomyolysis in patients who take this drug. This might represent a lethal dose, on a repeated basis, for a sensitive subgroup. One report estimates that 2% to 5% of 'statin' patients experience myopathy. This was said to be a low estimate because it is hard to establish this effect (generally based on the patient's symptoms) because muscle pain seems to be a common occurrence in people not taking the drug. Relevant, reported data are: (a) the frequency of rhabdomyosis is rare but increases with dosage and (b) muscle-related symptoms occurred with various regimens as follows: Fluvastatin 40 mg = 5.1%; Pravastatin 40 mg = 10.9%; Atorvastatin 40 to 80 mg = 14.9%; Simvastatin 40 to 80 mg = 18.2% (a data source is provided). It also is stated that while most people who experience muscle pain 'return to normal' some susceptible people suffer permanent muscle damage from statin use [8]. Also, the muscle break-down products can 'overload' the kidneys producing additional problems. More specifically, this site also states: "Severe rhabdomyolysis can result in death from acute kidney failure due to overload of the kidneys with deteriorated muscle tissue." This same report [8] cites studies from Denmark that report "… an individual who is a long-term user of statin drugs has anywhere from 4- to 14-times greater risk of developing peripheral neuropathy than a person who does not take statin drugs," As a bit of, perhaps, unneeded justification it can be recalled that snake venoms and other toxins whose envenomations have resulted in no documented deaths in humans, also are reported in this book.

Indeed, all drugs have unwanted effects and the dosage is controlled to manage these. Terms used for these properties of drugs are called: the therapeutic effect (or dose), the toxic effect (or dose), and the lethal effect (or dose) (Fig. 3). The effective dose and toxic dose have definitions corresponding to the LD_{50} that has been previously defined. Similarly, these values refer to doses affecting half the population treated with the drug. An ideal drug would have a therapeutic effect at a dose that has no toxic or lethal effect. Only serious, life-threatening conditions are treated with drugs that have a lethal effect at concentrations near the therapeutic dose. Cancer drugs are an example.

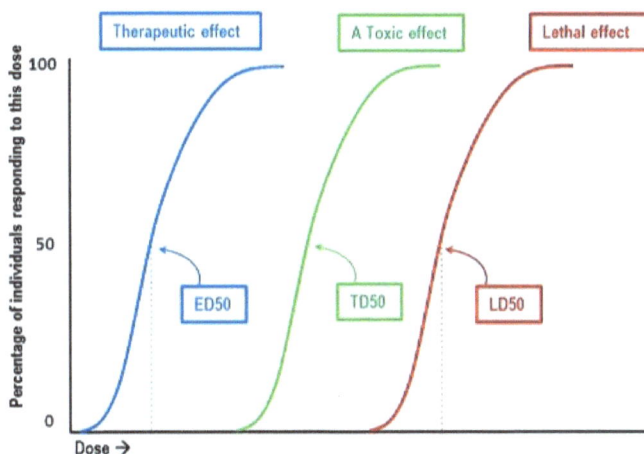

Fig. (3). Effects of drugs measured as effective dose for 50%, toxic dose for 50%, and lethal dose for 50%.

Cancer Drugs

A website alphabetically lists a large number of cancer drugs with summary information about how they work [9]. There also is a list of more than 200 cancer drugs with information summaries by the National Cancer institute [10]. This site describes the cancers treated by the drugs, research results, possible side effects, approval information, and ongoing clinical trial. All cancer treatments produce significant side effects and this site states: "Cancer treatments can cause side effects – problems that occur when treatment affects healthy tissues or organs" and there is a list of 23 common side effects. The pain and intrusion into daily life by these drugs is challenging to hear about and they must be extremely challenging for those who must take these drugs. Types of cancer treatment are also described and range from one drug to a combination of drugs for most treatments which may include surgery with chemotherapy or radiation therapy. Newer techniques include immunotherapy and targeted therapy which seeks to respond to changes in resistance of tumors that occur as treatment continues. Also, there is hormone therapy, stem cell transplants, and precision medicine, a therapy which seeks to use information gained from genetics to direct therapy.

It is significant that cancer chemotherapy requires stopping or slowing the growth of cancer cells to cause their death. Cancer cells grow and divide quickly, relative to normal cells, and drugs are designed to take advantage of this difference to kill cancer cells while sparing normal, healthy cells. This is a challenge since most agents powerful enough to kill fast-growing cancer cells, will also kill or slow the growth of healthy cells, especially those normal cell types with relatively higher division rates. Examples of fast-multiplying cells are those that line the inside of the mouth and intestines and those that cause hair growth. Damage by chemotherapeutics to healthy cells is the cause of side effects such as mouth sores, nausea, and hair loss. Side effects often get better or go away after chemotherapy is completed [10]. Because the list of cancer chemotherapeutics is very extensive, one will be described in some detail as an example with focus on its toxic side-effects.

Doxorubicin

Doxorubicin (trade name Adriamycin®) is widely used alone or with other drugs to treat many types of cancer including: acute lymphoblastic leukemia, acute myeloid leukemia, breast cancer, gastric cancer, Hodgkin lymphoma, neuroblastoma, non-Hodgkin lymphoma, ovarian cancer, small cell lung cancer, soft tissue and bone sarcomas, thyroid cancer, transitional cell bladder cancer, and Wilms tumor [10]. Doxorubicin is described as a 'frontline' drug for cancer and has been in use for more than 35 years. However, it causes toxicity to most organs and can cause life-threatening cardiotoxicity which mandates careful control of dose and duration of therapy.

Doxorubicin kills cancer cells and produces its serious side-effects because of its mechanism of action at the cell level (Fig. **4**). It binds to certain enzymes, intercalates into DNA, and has other molecular targets that affect both healthy cells and cancer cells. Several of these mechanisms are complex and a simplified description is necessary. A significant effect of doxorubicin is activation of cell signaling that induces apoptosis, a process of programmed cell death. Unfortunately, the challenge is to kill cancer cells more than normal cells. Death in healthy tissue causes toxicity in the brain, liver, kidney and heart [11].

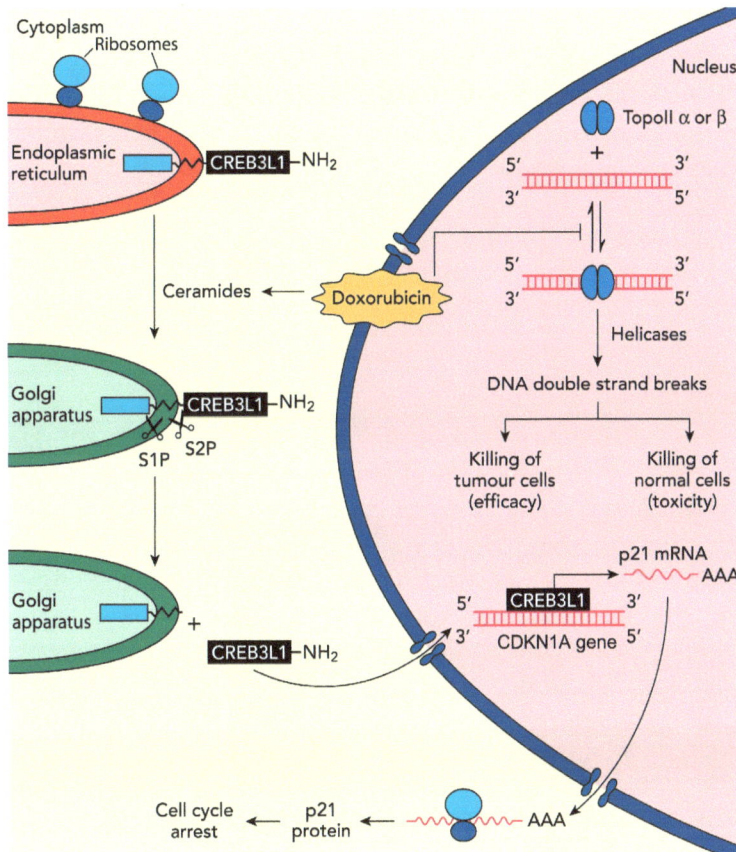

Fig. (4). Mechanisms of action of doxorubicin.

Research reported from my laboratory showed that doxorubicin decreases cardiac pyridine nucleotide coenzymes in a mouse model [12]. A major problem with most cancer chemotherapeutics is that they are not specifically selective for cancer cells and they will also, to a significant extent, eliminate normal, healthy cells.

Doxorubicin belongs to a class of cancer drugs that work by damaging cellular DNA in a way that prevents replication of DNA that is necessary for cell growth and division, a process called alkylation of DNA. Thus, doxorubicin specifically interferes with enzymes required for DNA replication. Unfortunately, its effect is not completely selective for cancer cells. More specifically, doxorubicin inhibits certain enzymes (topoisomerase I and II) required for DNA synthesis, and inserts itself into the structure of DNA in a manner that interferes with DNA replication. Doxorubicin 'intercalates' between base pairs in DNA's double helical structure. Subsequently, affected cells are 'programmed' for cell death. This process, called apoptosis, is triggered when the attempt by the cell to repair the damaged DNA fails and cell growth and division stops at particular stages of cell division. Ultimately, in such cells, DNA replication ceases as does RNA transcription and thus protein synthesis. An addition cellular mechanism of action leads to additional side-effects that are unwanted. Doxorubicin also affects cell membranes (both cancer cells and normal cells) by directly binding to proteins in the cell membrane with a consequence that doxorubicin is reduced chemically (gains electrons). Subsequently, the reduced doxorubicin can pass these electrons on to oxygen with the formation of reactive free radicals. These free radicals, in excess, are toxic to cells.

The toxicity of doxorubicin thus causes serious side effects including cardiotoxicity. Other common side effects include acute vomiting and nausea, baldness, light-headedness and even hallucinations [11]. Cardiotoxicity increases with total accumulated dose of doxorubicin and when it reached 500 mg/m^2 the risk of cardiomyopathy is significant and congestive heart failure occurs in approximately 20% of treated patients [11]. Cardiotoxicity is mediated at the molecular level by oxygen radicals (an area of my research) previously mentioned. Doxorubicin toxicity has been assessed in mice weighing 20 grams by injection into the tail veins using 10 mice per dose. The LD_{50} of doxorubicin is 14.2 mg/kg (two doses were given) [13]. Cyanide is only approximately 5-times as toxic. This is remarkable and helps us understand that very toxic drugs must be used when a life-ending disease like cancer is to be treated.

STRYCHNINE

Strychnine was once a medicinal listed in the *British Pharmacopoeia*, because it 'invigorates' nerves, but it is no longer listed because of its harmful effects. Strychnine affects chemical receptors on motor neurons – nerves connect with muscles to control movement. The marathon race (26.2 miles) requires a lot of muscle contractions. The 1904 Olympic marathon had many unusual features, including that the winner (Thomas Hicks) was given strychnine [14]. Strychnine is a powerful stimulant of motor nerves and produces contractions. At a dose of approximately 100 mg it can cause whole body convulsions and can kill by paralyzing breathing muscles; the lowest known lethal dose is reported to be only 36 mg [14]. At low doses strychnine is a stimulant.

There are various accounts of the 1904 marathon; most are similar and all include stories about the use of strychnine. The following is taken primarily from the account by Kathryn Harkup [14]. The race was run in St. Louis in 90 degree heat with no water stops. The top two runners were Frederick Lorz and Thomas Hicks. Lorz was a bricklayer by trade and Hicks was a brass worker. Lorz dropped out at the 9th mile, and it is said that he was driven by car to the stadium (the finish of the race) to collect his things. The car broke down at the nineteenth mile and Lorz decided to jog the rest of the way. He was first to cross the finish line but was disqualified when the facts were learned. It is said that Lorz claimed it was a joke. The winner, Hicks, is where strychnine comes in. He took an early lead but soon was struggling. Photos taken during the race show Hicks being supported under both arms by two men. Hicks was given approximately 1 mg of strychnine sulphate and some brandy. This temporarily revived him and he later was given a second dose of strychnine, presumably the same amount. He crossed the finish line but collapsed and was too weak to receive his medal. It is said that he recovered and lived until the age of 76 but never competed again. Strychnine was not a banned substance at the time (nor was brandy).

I can report from personal experiences that my fellow competitors in marathons and other long races are good guys but sometime they have strange quirks. Besides marathons, I recall entering a 100-mile walk which was held on a quarter-mile track around a football field. Surely, one has to be a little outside the norm to complete 400 laps within a 24-hour time limit. I recall one competitor consuming a can of spaghetti and meatballs during the event but never ceasing to walk. I have run marathons in Missouri and can testify to the difficulty imposed by terrain and temperature.

The tale of another competitor in this race (Felix Carbajal) speaks to the unusual athletes in this race. Carbajal was a Cuban postman and had lost his money for travel to the Olympics in a dice game in New Orleans. He hitchhiked to St. Louis and competed in street clothes with the trousers cut off at the knees (presumably to make himself more presentable as if wearing shorts). It is said that Carbajal started well but needed to take a break on the very hilly, dusty course. Carbajal stopped in an orchard to eat some apples which disagreed with him so he took a nap to recover and still finished fourth!

The 1904 Olympics, America's first, were tied to the World's Fair and the story told above is similarly recounted in a web page by Karen Abbott [15]. This site also reports that the marathon competitors were an odd assortment including a few who had run marathons, but "… the majority of the field was composed of middle-distance runners and assorted oddities." Fred Lorenz did all his training at night because his day job was as a bricklayer and had earned his spot in the Olympics by placing in a 'special five-mile race'. There were 10 Greeks who had never run a marathon, two men of the Tsuana tribe of South Africa who were said to have arrived at the starting line barefooted and, of course, Carbajal who was previously mentioned. Carbajal was only 5 feet tall and he was "… attired in a white, long-sleeved shirt, long, dark pants, a beret and a pair of street shoes. A fellow Olympian took pity, found a pair of scissors and cut Carbajal's trousers at the knee".

The reason strychnine was used by athletes and promoted by their trainers stems from how strychnine works biochemically. As long ago as 1933, considerable was known about this poison [16]. This review states: "Strychnine is the outstanding representative of the so-called central-convulsant poisons … The typical early symptoms of poisoning with a sufficient dose of strychnine are a marked hyper-flexia, soon to be followed by generalized convulsions due to unbalanced, maximal contractions of practically all striped muscles." Strychnine is known to cause muscle contraction, but not in a medically beneficial way [17]. Modern science has discovered that strychnine interferes with the normal process by which electrical signals are sent across junctions between nerve cells and muscle fibers which are caused to contract. For the message to pass this junction, a chemical signal is required. Actually, two chemical neurotransmitters are involved: acetylcholine which carries the impulse across the junction, and causes muscle contraction, and glycine which cause contraction to stop. A balance of these two signals is required to allow smooth contraction and relaxation of muscles. Strychnine attaches to glycine receptors and blocks the 'stop' signals. In this condition, the slightest of nerve stimulus results in continued and strong contraction. Strychnine binds 300- times more strongly to the receptors than does glycine [18].

Strychnine poisoning begins with muscle twitching and progresses rapidly (if sufficient strychnine is present) to horrific, whole body convulsions. The muscles of the back are usually stronger than abdominal muscles and the imbalance causes arching upwards with arms clinched at their sides and only the heels and back of the head of the wretched victims touching the ground or floor. These spasms recur as a series of convulsions. "Most victims survive between two and five of these bouts before the muscles controlling their breathing become paralyzed and they suffocate… During the whole process the victim is horribly and painfully aware of what is going on, as the nerves of the brain are also stimulated to give heightened perception. It is an agonizing way to die. Thankfully, strychnine rarely causes death… a paper published in 1979 explains that in the preceding 30 years there was on average one death per year in the UK due to strychnine poisoning; all but one were suicides… It is now illegal to purchase strychnine in the UK" [17].

Strychnine fits my definition, along with arsenic and poison hemlock, of a classical poison and it was available from the late 18th centuries. Through the late 19th century it was found in tonics now considered quackery. Its availability and uses to kill rats and sometimes coyotes and other wild animals, led to generalized knowledge of its toxicity. Strychnine is a bitter alkaloid in the seeds of a tree, *Strychnos nux-vomica,* (Fig. **5**) said to have been introduced into Europe in the 1500's from Asia [18]. French chemists, Joseph Bienaimé Caventou [1795-1877] and Pierre-Joseph Pelletier [1788-1842] isolated strychnine in 1818 and these two scientists were co-discovers of quinine and caffeine as well. Strychnine was synthesized in the laboratory in 1954 by Robert Woodward [1917-1979] [18]. Robert Woodward is considered by many to be the preeminent organic chemist of the 20th century.

Fig. (5). Strychnine tree showing the seeds which are the source of strychnine.

Strychnine is a good example of a poison that has had many uses with dose being the characteristic that determines the application, but in this case the prior uses have fallen away and even its use as a rodenticide has given way to safer chemicals.

CURARE

Curare is a classic example of the type of antagonists known as nicotinic Acetylcholine nicotinic receptor antagonists. It affects chemical transmission of electrical impulses across neuronal junctions (Fig. **6**). By competing with acetylcholine (the normal transmitter) it is an effective blocking agent for general

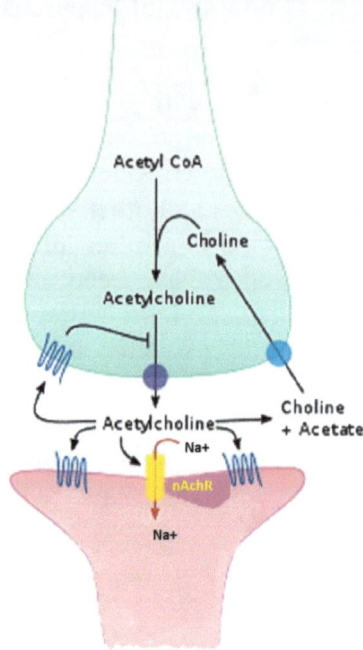

Fig. (6). Curare blocks transmission of impulses across myoneural junctions.

anesthesia. It is also described as a non-depolarizing blocking agent [19]. This site [19] has much information about curare and provides references to original sources. It describes a variety of related non-depolarizing blocking agents and rocuronium has the fastest onset (within 1 minute) and maximum effect within 3 minutes at a dose of 1.2 mg/kg.

Tubocurarine is the active form of curare which is found in extracts of the South American plant of the genus *Chondodendron* (Fig. **7**). Crude extracts of this plant were traditionally used by certain South American natives, and purified tubocurarine was once used as a skeletal muscle relaxant for surgery but has been replaced by synthetic agents that mimic the action of curare [19]. Tubocurarine does not pass through the blood-brain barrier and peripheral application of this agent results in selective paralysis of striated skeletal muscle. The LD_{50} in mice is 0.63 mg of agent per kg of mouse [19].

Fig. (7). Curare plant, the source of curare and d-tubocurarine.

Some details about the site of action of curare may be helpful. There are two main classes of receptors for acetylcholine (whose function was briefly described above): muscarinic receptors which are activated by muscarine and blocked by atropine, and nicotinic receptors which are activated by nicotine and blocked by curare. There are subtypes of these receptors. Importantly for understanding curare, only nicotinic acetylcholine receptors are present at the motor endplate of vertebrate skeletal muscles. Muscarinic receptors are found on smooth muscles and gland cells and together with nicotinic receptors, on autonomic ganglion cells and neurons in the central nervous system.

The botanical origin of curare as an arrow poison of South American Indians is of historical interest. In particular, Alexander von Humbolt [1769-1859] (who is said to have travelled 24,000 miles in attempting to understand the relationships between nature and habitat) described the process of making poisoned arrows [19]. An 'old Indian' was the community chemist. In his 'laboratory' he boiled plant sap, evaporated it to condense the toxin, and filtered it. Arrow tips were dipped into this thickened poison which was solution and not dried toxin; thus, the poison entered the system of its victim more rapidly and produced almost instant paralysis.

An account of the effects of curare on a human is instructive [20]. In 1946, a man named Frederick Prescott was paralyzed. He was lying on a table and completely unable to move. Two minutes before he had been injected with 30 mg of d-tubo-curarine a purified chemical from the curare plant. He was unable to open his eyes, he could not speak or swallow and no one in the room was aware of his distress. Within another minute his breathing became rapid and shallow and he later described himself as terrified. Just as he lost consciousness, his colleagues noticed his distress and started artificial respiration and manual compression of his chest. It took seven minutes for him to breathe on his own again and another thirty minutes to regain use of his body. Six hours later he was completely recovered. Prescott is said to have stated: "To be conscious yet paralyzed and unable to breathe is a very unpleasant experience" [20]. Dr. Prescott was Research Director of the Wellcome Research Institute and he had volunteered to try this agent on himself as an anesthetic. Indeed, there was a deep need for a better anesthetic than those available for abdominal and thoracic surgeries: nitrous oxide, cyclopropane, and ether. It was desirable to have an agent that is a complete muscle relaxant but one that does not produce shock and vomiting. Tubocurarine was a significant advancement but it did not block pain; it produced anesthesia without analgesia; it caused unconsciousness but not freedom from pain; therefore, its usefulness in surgery was limited.

The site that describes Dr. Prescott's experiences [20] also gives information about Alonso Perez de Tolosa [ca. 1490- 1549] and his encounter in the country now known as Venezuela with a peaceful tribe call the Bobures. The account states: "[They] fought just with blowguns, into which they put tiny arrows, dipped in poison. These, even if they cause only a slight wound, deprived the victim of his senses for two or three hours, the time it took the Indians to flee, and afterwards he recovered without further hurt (a reference is given)". It seems there were three forms of arrow poison: poison stored in a tube (tube poison), stored in a gourd (calabash poison), and stored in an earthen pot (pot poison). The most toxic was tube curare and it was typically prepared from the bark of the jungle plant *Strychnos toxifera* (a reference is provided). This plant is related to the plant from which strychnine is obtained as previously noted.

Thus, curare poisons sites that control peripheral nerves and toxicologists are taught the mnemonic SLUDGE to recall the effects of these agents: salivation, lachrymation, urination, defecation, GI upset, and emesis. But there are also effects on voluntary movement controlled by the nervous system.

PESTICIDES

Toxicologists lump the poisons that are used to control unwanted plants, insects, and rodents into the general category pesticide, but each has its own special category: herbicide, insecticide, and rodenticide, respectively. Examples will be provided for three agents of the huge number available. Indeed, most of the agents that we use to control plants, and some that are effective for insects, have been especially selected not to affect humans because they poison intracellular mechanisms that are absent in humans. Three pesticides that have significant toxicity for humans are paraquat, warfarin and the similar chemical Brodifacoum. Paraquat is chosen because of my personal interest and the opportunity to describe a mechanism that involves toxic oxygen radicals and the other two because they have an effect that is similar and which involves interference with blood clotting. We have encountered this before with certain toxins in snakes (Chapter 8) and this provides the opportunity to further examine the biochemistry of blood clotting which is intricate and shows evidence of extreme design.

Paraquat

Paraquat ingestion is said to be a leading cause of fatal poisoning in many parts of Asia, Pacific nations, and the Americas. Self poisoning is a common method of suicide and the magnitude of the problem was investigated in research published in 2007 [21]. About one-third of the World's suicides were said to have been committed with pesticides. These researchers specifically stated: "We conservatively estimate that there are 258,234 (plausible range 233,997 to 325,907) deaths from pesticide self-poisoning worldwide each year, accounting for 30%... of suicides globally... The proportion of all suicides using pesticides varies from 4% in the European region to over 50% in the Western Pacific region." This site gives a detailed breakdown of suicides by region and by country and provides references. In the U.S. pesticides are rarely used in acts of self-harm and there were only 87 suicides from 1995-1998 (approximately 22 per year). Data on this site included: "Acts of self-poisoning are associated with a case fatality of between 1% and 70%, depending on the particular pesticide taken. Paraquat and aluminum phosphide self-poisoning have case-fatalities in excess of 70% whereas case fatality for the organophosphorus (OP) insecticides dimethoate and chlorpyrifos are 23% and 8% respectively (references are given). This shows that paraquat is a significant poison and that its toxicity is high. Using Ovid Medline and combining the terms Paraquat and LD_{50}, a total of 57 papers were retrieved (11-30-2017). Most of these reports were for non-human data, of course.

In personal research, the effects of paraquat on bacteria and on laboratory animals were studied. Paraquat is redox-active and generates oxygen radicals during aerobic cell metabolism. Some aspects of paraquat poison (effects on the lungs, for example) are similar to the effects of exposure to hyperbaric oxygen. Sites of impairment in common for aerobic paraquat and oxygen poisoning in the bacterium *Escherichia coli* include: biosynthesis of certain amino acids, induction of genetic stringency, decreased content of thiamine, and impaired biosynthesis of pyrimidine nucleotides [22]. In other work from my laboratory, published in Science, the vitamin niacin was found to reduce the toxicity of paraquat for rats [23]. Male Sprague-Dawley rats, averaging approximately 200 grams each

(25 rats per group) were given two doses, intraperitoneally, of paraquat at 30 mg paraquat per kg body weight, 24 hours apart. Rats began to die after approximately 30 hours and 50% were dead by 60 hours. Rats that received the same paraquat treatment but also 500 mg of niacin per kilogram of body weight every 24 hours for 5 days beginning with the first administration of paraquat, began to die almost 36 hours later and did not reach 50% mortality until approximately 120 hours and the differences were significant at the 0.05 probability level. These experiments were done in 1981 when toxicity experiments with rats were still considered necessary and appropriate. Today, experiments of this type are no longer, or rarely, done. During this period (1982), I was serving as an unpaid consultant to Dr. Carlton Turner, Drug abuse Policy Office, President Ronald Reagan, the White house. Paraquat was believed to have caused more than 400 human deaths (prior to 1977) from accidental and suicidal ingestions, and perhaps of more current significance, paraquat was being sprayed on marijuana in Mexico and this was of concern.

Paraquat, unlike most herbicides, has significant toxicity for humans. A single dose of paraquat, given intraperitoneally at 50 mg per kg in mice, was reported in a study done in 2002 to induce lung toxicity [24]. Paraquat at a dose of 79 mg per kg was lethal for 50% of Sprague-Dawley or Wistar strains of rats within 24 hours as reported in a study done in 1996 [25].

Warfarin and Brodifacoum

I chose to include the Coumadin derivatives warfarin and brodificon in this chapter as examples of agents that have a poisonous effect by reducing the ability of blood to clot. A similar biological effect was described in this book for various biological toxins, especially certain snake venoms (Chapter 8). Other reasons for inclusion are the uses of warfarin and brodificon as rodenticides and warfarin's medical use as a blood thinner.

Warfarin is the active ingredient in several commercial rodenticides. It was the first anticoagulant used as a rat and mouse killer and was introduced in 1952 [26]. This site "Extoxnet Warfarin" is provided by The Pesticide Information Project which involves Cornell University, Michigan State University, Oregon State University, and the University of California at Davis [26]. Warfarin is odorless and tasteless and effective at very low doses. It is slow-acting and it is said that about a week is required to produce a significant effect on populations of rats and mice. Rodents tend to become bait-shy toward poisoned substances and this does not occur with warfarin. Death occurs by internal bleeding. This web site has an extensive review of the toxicity of warfarin and includes references. The LD_{50} for warfarin (acute oral dose) has been reported variously as: 3 mg/kg, 1.6 mg/kg,

186 mg/kg, 58 mg/kg, and the acute oral dose for rats over 4-5 days was reported to be 1.0 mg per kg. This wide variation is due to various factors and is typical of LD_{50} values reported historically from different laboratories. For warfarin, part of the variability appears to be that male rats and female rats respond differently with LD_{50s} of 323 m per kg for males and 58 mg per kg for females. This site also states that "On a multiple-dose basis, the reported LD_{100} is 0.2 mg per kg per day for 5 days (LD_{100} is 100% deaths). This site also reports toxicity values for various other domestic animals. It also states an oral-woman TDlo of 15 mg per kg per 21 weeks intermittent; 10.2 mg per kg oral-man TDlo (no time given). TDlo refers to the lowest dose that produces signs of toxicity. These values are provided here as an indication of the difficulty in determining precisely how much of any given substance is required to be recognized as toxic or to kill.

This site [26] also provides authoritative information about the toxic effects of warfarin. "Animals killed by warfarin exhibit extreme pallor of the skin, muscles, and all the viscera... evidence of hemorrhage may be found in any part of the body but usually only in one location in a single autopsy. Such blood as remains in the heart and vessels is grossly thin and forms a poor clot or no clot... Symptoms of human exposure to warfarin include hematuria, back pain, hematoma in arms and legs, bleeding lips, mucous membrane hemorrhage, abdominal pain, vomiting, and fecal blood... One source [reference provided] stated that serious illness was induced by the ingestion of 1.7 mg of warfarin per kg per day for 6 consecutive days with suicidal intent... All signs and symptoms were caused by hemorrhage and, following multiple transfusions and massive doses of vitamin K, recovery was complete."

To understand the site and mechanism of toxicity of warfarin, it is necessary to examine the mechanisms underlying the clotting of blood. It is essential to life that blood be able to clot, but that clotting not occur except under certain conditions. Some of the complex factors required for clotting of blood and for control of clotting are shown in Fig. (**8**). It is convenient to think of blood clotting as occurring in stages sometimes called separate pathways [27]. One pathway component is commonly called intrinsic (red color in the figure); another part is called extrinsic (grey color in the figure); and another part (blue color in the figure) is called common. The figure also shows the pathway of clot dissolution (light green color in the figure). The extrinsic pathway is activated by external trauma that causes blood loss from the blood vessels. The intrinsic pathway refers to the components necessary for clotting but which are in an inactive form in blood. Blood coagulation is the process by which blood clots to prevent bleeding. Stopping blood loss is an exquisitely controlled process. In the simplest of terms, vasoconstriction is the first response to injury to the vascular wall. Reduced blood flow to the injured site conserves blood in the circulatory system. Platelets (a special type of blood cells) aggregate at the injury site and from a sticky plug if the injury is small. Platelets also initiate clotting of fibrin. Clot formation is complex and tightly controlled. Several factors, called clotting factors, are present in blood in forms that can be activated sequentially by what is known as a 'cascade'. This results in exquisite control of the process. Soluble plasma protein, fibrinogen, is converted into fibrin an insoluble plasma protein which is a primary constituent of the blood clot. In the extrinsic pathway, activation of a clot results from trauma that causes blood loss. This pathway is more rapid, for safety purposes, than is the intrinsic pathway. The intrinsic pathway is activated by trauma inside the vascular system. Platelets, exposed endothelium, chemicals and collagen are involved and the pathway is slower. There is also the common pathway where intrinsic and extrinsic pathways converge to complete the clotting process.

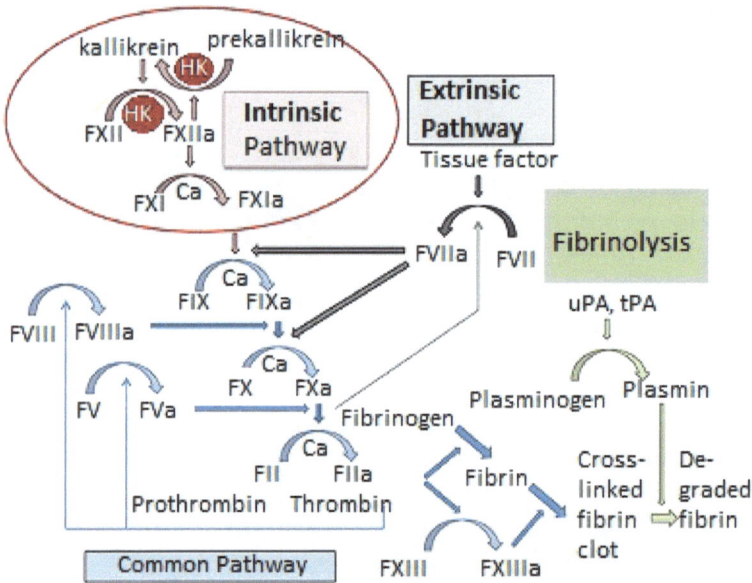

Fig. (8). Blood clotting pathways and mechanism; see text for details.

Because blood clotting and its control are essential physiologically and for understanding the mechanism and site of action of warfarin (and Brodifacoum also), a second figure (Fig. **9**) describes the clotting of blood in a slightly simpler manner.

Brodifacoum is a compound similar in structure to warfarin (Coumadin® is a brand name for warfarin). It is sometimes called 'second generation' or 'superwarfarin'. It has a longer duration of action that warfarin because its elimination from the body is much slower. It specifically inhibits the enzyme vitamin K epoxide reductase and it steadily decreases the concentration of active vitamin K in the blood. Vitamin K is required for proper clotting of blood. In addition, Brodifacoum increases permeability of capillaries and leakage occurs into tissue. Poisoning causes internal bleeding leading to shock and can result in death. The oral LD_{50} for the mouse is reported to be 40 mg per kg [28].

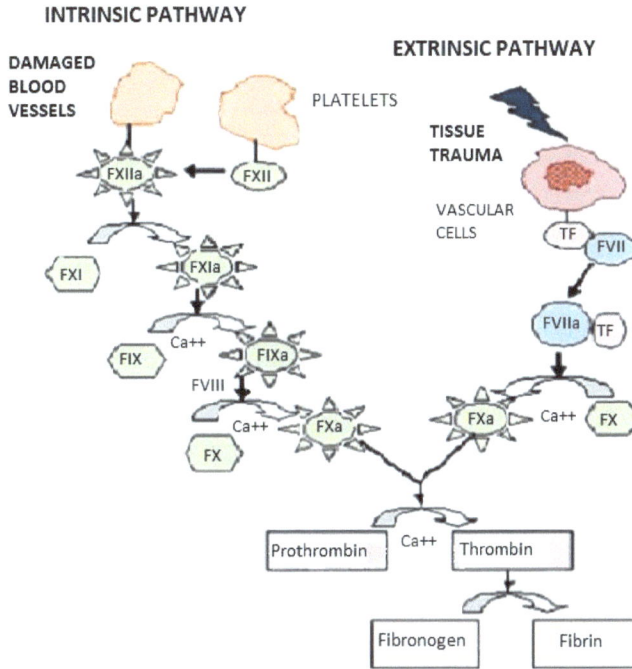

Fig. (9). Blood clotting somewhat simplified.

Warfarin (and brodifacoum) poison by virtue of interfering with blood clotting mechanism (Fig. **10**). In this figure the following definitions may be helpful. R-warfarin and S-warfarin refer to enantiomorphs which are present in equal amounts naturally in warfarin samples as slightly different shaped molecules which are recognized differently by liver enzymes, shown with CPY followed by other alpha-numerics. These enzymes modify warfarin by adding the OH (hydroxyl) functional group in various places on the warfarin molecule and the significant point is that the hydroxylated forms of warfarin are essentially inactive and are excreted. Thus, the liver has this mechanism of detoxifying warfarin. The comparative abundances of these liver detoxifying enzymes in individual patients is important clinically in the use of warfarin as therapy. Both forms of warfarin block the enzyme (shown in the figure as VKORC1) that converts oxidized vitamin K into reduced vitamin K. It is this reduced form of vitamin K that is required for activating the blood clotting factors shown at the bottom left of the figure into activated factors, shown at the bottom right. A fuller description of these processes may be helpful and is available at [29].

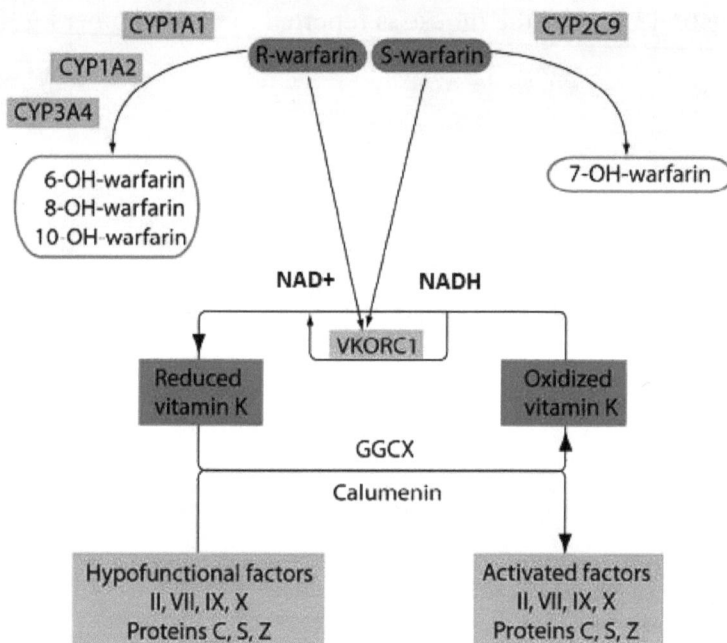

Fig. (10). The site of action of warfarin. Details of modification of warfarin by liver enzymes (CYP) are shown. NAD$^+$ and NADH are oxidized and reduced forms, respectively, of nicotinamide adenine dinucleotide; GGCX and VKORC1 are enzymes; see text for details.

An excellent source of information about vitamin K in health and disease, including its role in coagulation, is found in the website of the Linus Pauling Institute, Oregon State University [30]. The human has limited capacity to store vitamin K and this vitamin participates in an oxidation-reduction cycle to reuse it multiple times. Vitamin K is essential as a cofactor for adding a carboxyl group to a particular amino acid, glutamic acid, found in proteins and of particular relevance her since they are involved in blood clotting. Vitamin K deficiency increases the risk of excessive bleeding and vitamin K is injected in newborns to protect them from life-threatening bleeding within the skull. The choice of K as the designation for this vitamin stems from the German word *koagulation* and is related to its role in blood clotting which was known at the time of its characterization as a vitamin [30]. As stated, vitamin k is a cofactor for the enzyme ɤ-glutamylcarboxylase (GGCX) which catalyzes the addition of a carboxyl group to glutamic acid. This enzyme action occurs only on specific glutamic acid residues in certain vitamin K-dependent proteins and this is critical for their ability to bind calcium which is necessary for blood clotting. Vitamin K is a structural analogue of warfarin. The LD_{50} of Warfarin as listed by this site (National Toxicology Program of the NIH) is 8.7 mg per kg [31].

Alcohols

Methanol, ethanol, isopropanol, and ethylene glycol are similar in chemical structure but have different toxicities because of the way they are metabolized by humans, primarily by liver enzymes. These alcohols are metabolized by the enzyme alcohol dehydrogenase and then by other enzymes to form toxic metabolites (Fig. **11**). This chapter will focus on the toxic mechanisms and significance of these alcohols for human poisonings.

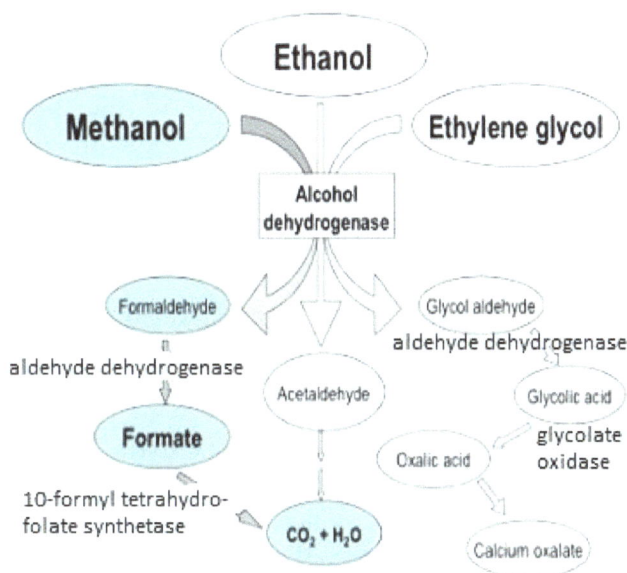

Fig. (11). Metabolism of the alcohols to form more toxic products.

The interesting site [31], previously cited, compares the toxicity (LD_{50}) for various agents. Ethanol is 14,000 mg, table salt is 3,000 mg, aspirin is 1,000 mg, caffeine is 130-320 mg, arsenic is 46 mg, and nicotine is 1 mg. These data indicate the low toxicity, with regards to lethality, for ethanol which is the alcohol in beer, wine and distilled spirits. With respect to ethanol, our interest here is on the acute toxicity and not on its addictive qualities, its social uses, and not even on its role in automobile accidents.

Ethanol is safely consumed regularly in moderation by many people. However, when consumed in excess, it can be deadly. Alcohol poisoning occurs [32] "... when there is so much alcohol in the bloodstream that all areas of the brain controlling basic life-support functions – such as breathing, heart rate, and temperature control – begin to shut down. Symptoms of alcohol poisoning include confusion; difficulty remaining conscious; vomiting; seizures; trouble with breathing; slow heart rate; clammy skin; dulled responses, such as no gag reflex (the gag reflex prevents chocking); and extremely low body temperature... Even if the drinker survives, an alcohol overdose can lead to long-lasting brain damage."

The question of what is a lethal dose of alcohol is made difficult to answer because there are wide differences in people. However, significant risk of death in most drinkers due to suppression of vital functions occurs in the range of 0.31 to 0.45% blood alcohol content (BAC) [32]; this site says it is life threatening. The LD_{50} for ethanol of 14,000 mg per kg of body weight stated elsewhere [31]. Another site [33] states that the lethal dose of alcohol is 13 shots where I shot is defined as 45 ml of 40% alcohol; therefore, the LD_{50} calculates to be approximately 2,670 mg per kg. Another site states the lethal dose in terms of blood alcohol level and gives the figure of 0.40% [34]. Another site [35], a material safety data sheet gives the following data for ethanol: "Oral (LD50): Acute: 7060 mg/kg [Rat]. 3450 mg/kg [Mouse]." If an overdosed victim arrives at a hospital, medical staff will need to manage breathing problems, administer fluids to combat dehydration, and low blood sugar, and flushing the drinker's stomach may be useful if alcohol is still being absorbed.

Ethyl alcohol (ethanol, drinking alcohol) is a simple, low molecular weight compound containing only three carbons, six hydrogens, and one oxygen. In addition to being the intoxicating agent in beer, wine and distilled spirits, ethanol is an ingredient in various food products, in cough and cold medications, and in mouthwashes. Ethanol is rapidly absorbed from the stomach and small intestines. It is converted by three liver enzymes: a particular cytochrome P450 called isoenzyme CYP2E1 found in cellular microsomes, an enzyme found in the cytosol of cells called alcohol dehydrogenase, and catalase found in peroxisomes [36]. People differ genetically in the rate at which they metabolize alcohol. It is alcohol itself and not its metabolites that produce the sought after and inebriating effects of alcohol. During the absorption phase, blood levels of alcohol rise and when drinking stops, blood alcohol levels fall.

Alcohol intoxication is a common problem in most societies today. "More than 8 million Americans are believed to be dependent on alcohol, and up to 15% of the population is considered at risk" [36]. This site also states: "Acute intoxication with any of the alcohols can result in respiratory depression, aspiration, hypotension, and cardiovascular collapse."

Ethanol is significantly toxic. "In 2014, 6026 single exposures to ethanol in beverages, with 219 major outcomes and 15 deaths, were reported to US Poison Control Centers" [36]. This site also reported: "There were 3508 non-beverage exposures with 13 major outcomes and 4 deaths. Ethanol-based hand sanitizers accounted for 18,322 single exposures, with 11 major outcomes and no deaths, and ethanol-containing mouthwashes accounted for 6,539 single exposures, with 20 major outcomes and two deaths (a reference is provided)".

Isopropanol (sometimes called rubbing alcohol) is used as a solvent and disinfectant and is found in mouthwashes, skin lotions, and hand sanitizers. It is profoundly intoxicating. Methanol is an industrial solvent and is often used as a paint remover. Methanol is primarily metabolized to formaldehyde and subsequently to formic acid and this is the cause of its toxicity for the eye where it can cause blindness. Ethylene glycol is odorless but is sweet-tasting. It is prominently found in antifreeze. It is metabolized to glycolic acid which is then converted in the body to oxalic acid which combines with calcium to form calcium oxalate crystals which can damage the kidneys.

Details are provided in the previously referenced source [36]. "In 2014, 15,334 single exposures to isopropanol… were reported to US Poison Control Centers... 60 patients were classified as experiencing 'major' morbidity, with one patient dying (a reference is cited). In the same year, 1,610 single exposures to methanol, including automotive products, and 5,552 exposures to ethylene glycol were reported... 16 patients were classified as experiencing 'major' disability, and 9 additional patients dying... these numbers likely underestimate the true incidence…". Details about alcohol poisoning deaths have been reported by the Centers for Disease Control and Prevention [37]. An average of 6 people die of alcohol poisoning each day in the U.S. (2010-2012 data), 76% are men, and there are 2,200 alcohol death poisonings each year (on average).

The primary toxicity with isopropanol is CNS depression and manifestations include lethargy, ataxia, and coma. This agent also irritates the GI tract causing abdominal pain, hemorrhagic gastritis and vomiting; however, unlike methanol and ethylene glycol, isopropanol does not cause metabolic acidosis [36]. Methanol toxicity causes severe acidic metabolic acidosis, and the formic acid produced metabolically is toxic. The eye is the primary site of organ toxicity but in the later stages of severe intoxication, damage to the basal ganglia also are significant. Pancreatitis can occur, and hyperventilation as a compensation to acidosis [36]. Ethylene glycol, itself, is practically non-toxic. However, the glyoxylic acid formed metabolically can cause acidosis and the oxalate crystals formed metabolically accumulate in the proximal renal tubules in the kidneys and can induce renal failure and death. Since oxalate combines with calcium, hypocalcaemia can ensue to cause coma, seizures, and dysrhythmias. Calcium oxalate crystals are deposited in the brain, heart, and lungs as well as the kidneys [36].

Several values for the LD_{50} for methanol with various administration routes are listed in a MSDS [38]: Values included intraperitoneal rat 7529 mg/kg; intravenous rat 2131 mg/kg; and oral rat 5600 mg/kg. The LC_{50} inhalation rat is 64,000 ppm with exposure of 4 hours. The immediately dangerous to life and health (IDLH) for humans is listed as 6,000 ppm.

The toxicity of ethylene glycol was reported as follows [39]: "Acute oral toxicity (LD50): 4700 mg/kg (Rat). Acute toxicity of the vapor (LC50): >200 mg/m³ 4 hours (rat), and Special Remarks on Toxicity to Animals: Lowest Published Toxic Dose/Conc.: TDL [man] – Route: oral; Dose 15 gm/kg Lethal Dose/Conc. 50% kill LD50 (Rabbit)- Route: dermal; Dose: 9530 μl/kg" [39].

Carbon Monoxide

Carbon monoxide is a gas and intoxication is by inhalation (Fig. **12**). Therefore, it is difficult to compare its toxicity with agents that are assessed after injection into the body or even with those that are commonly toxic by ingestion. I include carbon monoxide here because it is significantly toxic and it is toxic by an interesting mechanism – it binds tightly to hemoglobin and interferes with the transport of oxygen – and I choose to end this chapter by addressing the toxic nature of another similar molecule which I suspect may come as a surprise.

Fig. (12). Space-filling model of a molecule of carbon monoxide; carbon is shown in black and oxygen in red.

Carbon monoxide is present in the fumes from burning fuel in motor vehicles, wood-burning stoves, lanterns, grills, fireplaces, and gas stoves, ranges and furnaces. Carbon monoxide gas can accumulate in homes and reach toxic or even lethal levels. Carbon monoxide monitors are available to warn people from this danger. Common symptoms of poisoning by carbon monoxide are: dizziness, weakness, upset stomach, vomiting, chest pain, and confusion. Sometimes poisoning symptoms are described as 'Flu-like' [40]. This CDC site states: "Each year, more than 400 Americans die from unintentional CO [carbon monoxide] poisoning not linked to fires, more than 20,000 visit the emergency room, and more than 4,000 are hospitalized."

Carbon monoxide is especially dangerous because it not detectable by humans until serious symptoms begin to develop (Fig. **13**). It is true that small amounts of carbon monoxide are produced endogenously by the body as a product of the degradation of heme (a blood constituent) and carbon monoxide has been found to serve as a transmitter of molecular signals. In this function is like the other gases nitrous oxide and hydrogen sulfide and called a 'gaso-transmitter' [41]. The concentration of carbon monoxide, normally present in cells in the body, is miniscule and this is another example where: "the dose makes the poison".

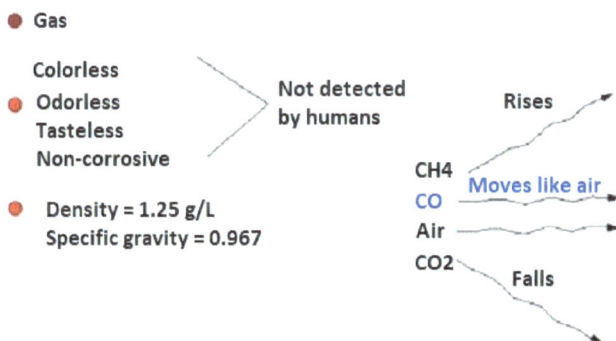

Fig. (13). Properties of carbon monoxide.

The toxicity of carbon dioxide gas in measured in terms of the lethal concentration in the breathing atmosphere that kills 50% of test subjects (LC_{50}) in a specified amount of time which must be stated. Thus, the LC_{50} is not equivalent to the LD_{50} which is based on a weighed, single, dose. It is possible to estimate the dose delivered by inhalation but it involves measuring or estimating the respiratory volume over the time the dose is delivered. A better estimate also involves a measurement or estimate of the absorbed dose, which is the effective amount of the toxic agent. The Material Safety Data Sheet, supplied by Matheson states for carbon monoxide a rat inhalation LC_{50} of 1807 ppm/4 hours [42].

The toxicity of carbon monoxide stems from the fact that carbon monoxide and oxygen bind to iron sites in hemoglobin in red blood cells (Fig. **14**). Oxygen is transported to all cells in the body as an essential molecule for life. Carbon monoxide is a toxin. It is toxic because it binds more avidly to hemoglobin than does oxygen. It displaces oxygen and is retained in red blood cells. Therefore, carbon monoxide interferes with oxygen transport by the blood and causes serious symptoms throughout the body (Fig. **15**). Tissues are starved for oxygen. Mitochondrial respiration of cells is specifically impaired and synthesis of ATP drops to a critical level (Fig. **16**). All cells in the body are affected but effects are seen most quickly in the brain.

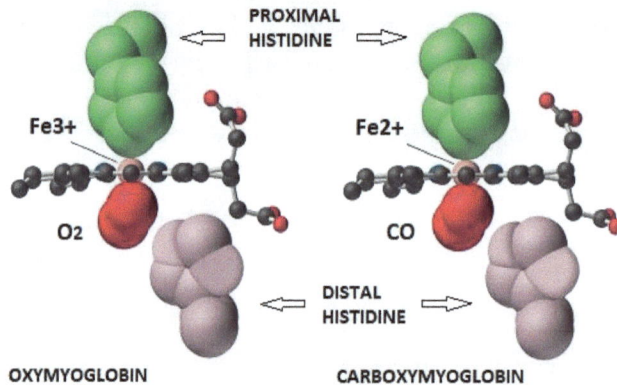

Fig. (14). Binding of oxygen and carbon monoxide to iron in hemoglobin in red blood cells. Two histidine (amino acid) sites near iron are binding sites for oxygen and carbon monoxide; however, carbon monoxide binds much more tightly. Even at low concentrations of carbon monoxide, oxygen transport is impaired.

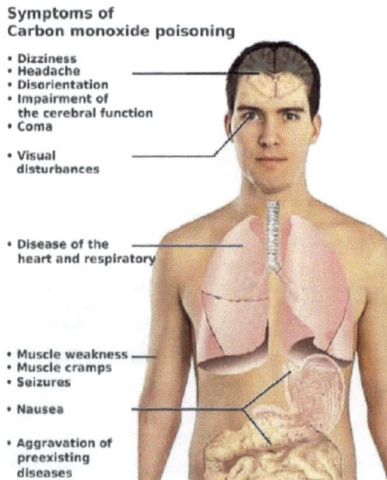

Fig. (15). Mechanism of carbon monoxide toxicity.

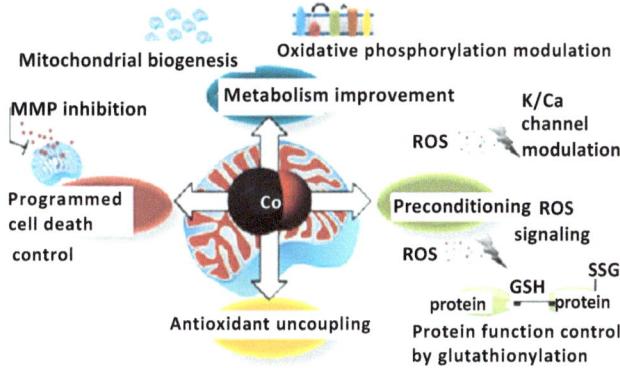

Fig. (16). Mitochondrial sites and mechanisms of action of carbon monoxide are dependent on the concentration of carbon monoxide and its site of action. At toxic concentrations the effect on oxidative phosphorylation can be lethal.

The effects of carbon monoxide at the level of the mitochondrion are complex (Fig. **15**). The effect on oxidative phosphorylation (ATP synthesis) has been mentioned. This alone is life-threatening when a lethal atmosphere of carbon monoxide is inhaled even briefly. At lower concentration there are effects on programmed cell death (apoptosis). At physiological levels, carbon monoxide serves as a cell signaling molecule which has been briefly described.

Oxygen

I expect that most readers will be surprised to find oxygen in this book of toxic agents. I recall many years ago when attending a meeting of scientists, I had the privilege of meeting and talking with Captain Albert R. Behnke [1903-1992]. He was even then the 'grand old man' of the U.S. Navy diving program. I was there to give a talk about the toxicity of oxygen. He approached me after my presentation and, grinning, he said words to the effect: "I am disappointed to hear you describe oxygen as toxic. Oxygen is one of the few positive things I can send down to my divers in the dangerous, cold, pressurized environment under the sea, and you are taking even that away!"

It is often said that people can live for a few weeks without food, a few days without water, but they can live only a few minutes without oxygen. A book I have written has the title "Oxygen, The Breath of Life: Boon and Bane in Human Health, Disease, and Therapy" [43]. In that book, I have described much of the history of research about oxygen including its toxic nature. In this chapter, only a brief account of the toxicity of this remarkable molecule, oxygen, will be given.

Oxygen at higher concentrations and partial pressures than are found in ambient air become toxic for cells through the body (Fig. **17**). Because oxygen is a gas, lethal exposure is measured as a lethal concentration not a lethal weight (dose). Technically, it is customary to measure the LC_{50}, the concentration which is lethal for the test subject, and the time of exposure must be specified. Although not customarily done, it is possible to measure the absorbed dose of oxygen and to specific this as the lethal dose. For this calculation it is necessary to know the oxygen uptake rate. This leads to a quandary. The consumption of oxygen by elite athletes has been studied, and a given athlete, at a given state of athletic training, will have a maximum rate at which oxygen can be consumed. Thus, the 'dose' of oxygen that a person at rest consumes, and the 'dose' of oxygen received by a person exercising at maximum rate, can be calculated. Neither of these doses of oxygen are acutely toxic, in the sense of 'poisoning' the cells of the body in a measurable way. It is conceded that some scientists believe that a lifetime is the time required for normal atmospheric oxygen (1 atm pressure and approximately 20% oxygen by volume) to be lethal. This is a rather extreme view and not provable.

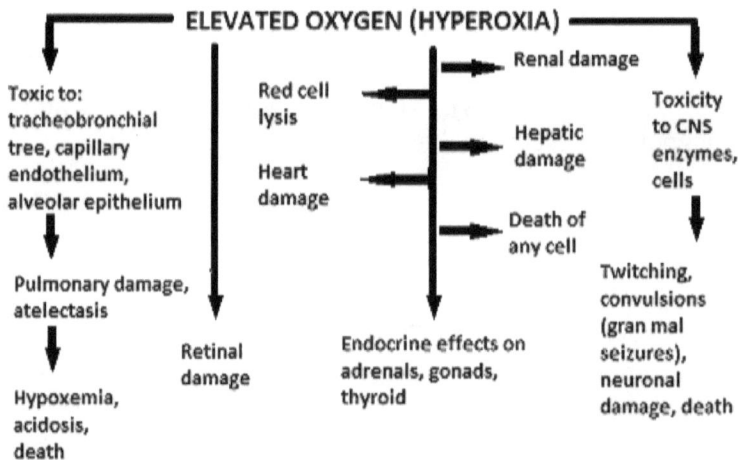

Fig. (17). Oxygen, when breathed at concentrations higher than that found in normal air, damages organs, tissues, cells, and subcellular components throughout the body.

Let us make some calculations and see where they lead. The average male (20-29 years of age) is reported to have a maximum oxygen uptake rate of 44-51 ml per kg body weight per min. This dose can be converted to the mass of oxygen. Let us use 50 ml per kg, for convenience. The oxygen consumed (dose) in 1 minute when breathing air is 71.4 mg oxygen. Lance Alworth (bicycle champion racer) has a much higher maximum oxygen uptake, 85 ml per Kg per min, which equates to 121.4 mg per min. We are sure that this is a safe dose, but it tells us nothing about the lethal dose.

There is no reliable information about the LC_{50} for oxygen for humans; no one has purposely exposed humans until they died in controlled experiments. In personal research, rats exposed to 6 atmospheres of pure oxygen convulsed an average of 4 times over 40 min, beginning at an average time of 21 min. Most rats did not survive. To convert animal dose to human, multiply by 0.16 (based on body surface area comparisons). However, this is for converting dose, not concentration, in the breathing atmosphere. Going further seems futile because it is likely to be very inaccurate. We do know that hyperbaric oxygen has greater lethal toxicity for the brain than for the lungs. Convulsions are surely a risk at 6 atmospheres oxygen.

Fig. (**18**) shows the effects of duration of breathing *vs.* inspired oxygen partial pressure and three decrements in lung vital capacities (a measure of lung impairment) are depicted by red, green and blue lines (curves). Breathing pure oxygen at 5 atmospheres (5.0 bars) begins to show measurable (2% decrement) in lung vital capacity within about 1 hour. An 8% decrement occurs in less than 4 hours and 20% occurs in about 6 hours (Fig. **18**). The toxicities of other durations and partial pressures of oxygen, as measured by effects on this measure of lung function, can be interpolated. Even comparatively low levels of lung impairment become important when oxygen therapy is maintained intermittently or continuously for long periods of time.

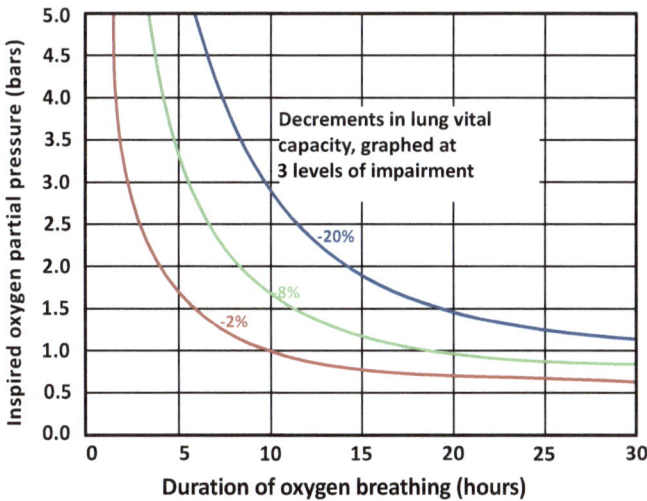

Fig. (18). Pulmonary oxygen toxicity (measured as decrement in lung vital capacity) and expressed at three levels: 2% (red), (8% green), and 20% (blue) plotted *versus* the oxygen partial pressure in the breathing gas (1.0 bar = 1 atmosphere pressure).

Interpretation of specific examples will be helpful for understanding Fig. (**19**). The 'Unimpaired Performance Zone' lies from only slightly below the approximately 20% oxygen found in ambient air to slightly less than 40% oxygen. For continuous, long-term breathing, an atmosphere containing 60% oxygen is not safe (Maximum Tolerance). Understanding these effects of oxygen is important for providing breathing atmospheres for deep-sea diving and for space capsules. It is also medically essential when providing either acute or long-term oxygen as a drug for various illnesses. And it becomes critical when oxygen therapy is administered in hyperbaric chambers for treatment of gas gangrene, carbon monoxide poisoning, and for other as yet experimental purposes.

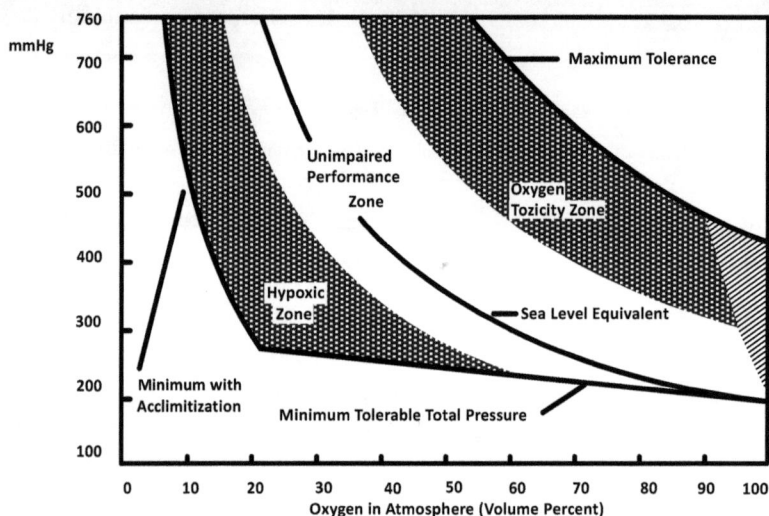

Fig. (19). The toxicity of oxygen in terms of several variables. Focus on the zones of Oxygen toxicity and maximum tolerance in terms of the volume percent oxygen in atmosphere (where normal air is approximately 20%) and total pressure in mm Hg (where normal air is approximately 760).

It is generally recognized that exposure to 1 atmosphere of pure oxygen begins to show some lung toxicity in about 10 hours, and below about 0.6 bar partial pressure of oxygen (about 60%), oxygen can be breathed safely for long periods of time. However, the zone of O_2 toxicity is of most interest. Look at the top horizontal line (Fig. **19**). The hatched portion of O_2 toxicity begins at only 40% oxygen. Note that this is only double the oxygen present in the normal atmosphere on earth. The maximum tolerated oxygen is only approximately 55%. It is remarkable to me that oxygen, essential to life, becomes harmful at approximately double that provided in normal air. Indeed, oxygen is classified as a drug by the Food and Drug Administration. Although it is not commonly reported, oxygen has a comparatively high therapeutic index (the ratio of amount that causes a therapeutic effect divided by the amount that causes toxicity). From the time of the discovery of oxygen almost 250 years ago near the time when our country gained independence, it has been known that oxygen was both required for life and potentially harmful. This paradoxical effect of oxygen results from unique aspects of atomic and molecular oxygen. These chemical effects derive from the arrangement of oxygen's electrons that permit a ready exchange of these electrons singly (one at a time) in various biologically-significant reactions. It is a blessing that oxygen has these special properties that permit complex life.

NOTES

[1] John Fletcher [1579-1625] Sicelides (c. 1614), Act V, Scene III.

REFERENCES

[1]   About the Epidemic | HHS.gov [Internet]. [cited 2017 Nov 29]. Available from: https://www.hhs.gov/opioids/about-the-epidemic/index.html

[2]   Zylla D, Steele G, Gupta P. A systematic review of the impact of pain on overall survival in patients with cancer. Support Care Cancer [Internet] Springer Berlin Heidelberg , 2017 May 11; [cited 2017 Nov 29];25(5): 1687-98. Available from: http://link.springer.com/10.1007/s00520-017-3614-y

[3]   Koo JW, Mazei-Robison MS, Chaudhury D, Juarez B, LaPlant Q, Ferguson D, *et al.* BDNF Is a Negative Modulator of Morphine Action. Science (80-) [Internet] , 2012 Oct 5; [cited 2017 Nov 29];338(6103): 124-8. Available from: http://www.sciencemag.org/cgi/doi/10.1126/science.1222265

[4]   Opioids | National Institute on Drug Abuse (NIDA) [Internet]. [cited 2017 Nov 29]. Available from: https://www.drugabuse.gov/drugs-abuse/opioids

[5]   Fentanyl Drug Profile [Internet]. [cited 2017 Nov 29]. Available from: http://www.emcdda.europa.eu/publications/drug-profiles/fentanyl

[6]   Fentanyl: What Is a Lethal Dose? [Internet]. [cited 2017 Nov 29]. Available from: https://www.oxfordtreatment.com/fentanyl/lethal-dose/

[7]   Statin | Definition of Statin by Merriam-Webster [Internet]. [cited 2017 Nov 29]. Available from: https://www.merriam-webster.com/dictionary/statin

[8]   Major Side Effects of Statin Drugs [Internet]. [cited 2017 Nov 29]. Available from: http://www.statinanswers.com/effects.htm

[9]   Cancer Drugs for Your Cancer Treatment | CTCA [Internet]. [cited 2017 Nov 29]. Available from: https://www.cancercenter.com/cancer-drugs/?source=BNGPS01&channel=paid

[10]   Cancer Drugs - National Cancer Institute [Internet]. [cited 2017 Nov 29]. Available from: https://www.cancer.gov/ about-cancer/treatment/drugs

[11]   Tacar O, Sriamornsak P, Dass CR. Doxorubicin: an update on anticancer molecular action, toxicity and novel drug delivery systems. J Pharm Pharmacol [Internet] Blackwell Publishing Ltd , 2013 Feb 1; [cited 2017 Nov 30];65(2): 157-70. http://doi.wiley.com/10.1111/j.2042-7158.2012.01567.x

[12]   Brown OR, Amash H, Perkins W, Schroeder R. Doxorubicin Decreases Cardiac Pyrimidine Nucleotide Coenzymes in the Bertazzoli Mouse Model. Med Sci Res 1987; 15: 497-8.

[13]   Deprez-De Campeneere D, Baurain R, Huybrechts M, Trouet A. Comparative study in mice of the toxicity, pharmacology, and therapeutic activity of daunorubicin-DNA and doxorubicin-DNA complexes. Cancer Chemother Pharmacol [Internet] Springer-Verlag , 1979 Nov 29; [cited 2017 Nov 29];2(1): 25-30. Available from: http://link.springer.com/10.1007/BF00253101

[14]   Harkup K. The Coctail of Poison and Brandy That Led to Olympic Gold [Internet]. Available from: https://www.theguardian.com/science/blog/2016/jul/21/the-cocktail-of-poison-and-brandy-tha--led-to-olympic-gold-strychnine

[15]   The 1904 Olympic Marathon May Have Been the Strangest Ever | History | Smithsonian [Internet]. [cited 2017 Nov 30]. Available from: https://www.smithsonianmag.com/history/the-1904-olymp-

c-marathon-may-have-been-the-strangest-ever-14910747/

[16] American Physiological Society (1887-) JGD de. Physiological reviews. [Internet]. Physiological Reviews American Physiological Society , 1933 [cited 2017 Nov 30];(): 325-35. Available from: http://physrev.physiology.org/content/13/3/325

[17] Strychnine: the notorious but rare poison at the heart of a modern mystery | Science | The Guardian [Internet]. [cited 2017 Nov 30]. Available from: https://www.theguardian.com/science/blog/2016/mar/17/strychnine-the-notorious-but-rare-poison-at-the-heart-of-a-modern-myster--saddleworth-moor

[18] Strychnine: Last of the Romantic Poisons | Nature's Poisons [Internet]. [cited 2017 Nov 30]. Available from: https://naturespoisons.com/2014/04/29/strychnine-last-of-the-romantic-poisons/

[19] Curare - an overview | ScienceDirect Topics [Internet]. [cited 2017 Nov 30]. Available from: https://www.sciencedirect.com/topics/neuroscience/curare

[20] Curare: From Paralyzed to Anesthetized | Nature's Poisons [Internet]. [cited 2017 Nov 25]. Available from: https://naturespoisons.com/2014/05/13/curare-from-paralyzed-to-anesthetized-tubocurarine/

[21] Gunnell D, Eddleston M, Phillips MR, Konradsen F. The global distribution of fatal pesticide self-poisoning: Systematic review. BMC Public Health [Internet] , 2007 Dec 21; [cited 2017 Nov 30];7(1): 357. Available from: http://bmcpublichealth.biomedcentral.com/articles/10.1186/1471-2458-7-357

[22] Brown OR, Seither RL. Oxygen and redox-active drugs: shared toxicity sites. Fundam Appl Toxicol 1983; 3(4): 209-14. [PMID: 6195038]

[23] Brown OR, Heitkamp M, Song CS. Niacin Reduces Paraquat Toxicity in Rats. Science [Internet] , 1981 Jun 26; [cited 2016 Jul 24];212(4502): 1510-2. Available from: http://www.ncbi.nlm.nih.gov/pubmed/7233236

[24] Mustafa A, Gado AM, Al-Shabanah OA, Al-Bekairi AM. Protective effect of aminoguanidine against paraquat-induced oxidative stress in the lung of mice. Comp Biochem Physiol Part C Toxicol Pharmacol [Internet] Elsevier , 2002 Jul 1; [cited 2017 Dec 1];132(3): 391-7. Available from: https://www.sciencedirect.com/science/article/pii/S1532045602000959

[25] Melchiorri D, Reiter RJ, Sewerynek E, Hara M, Chen L, Nisticò G. Paraquat toxicity and oxidative damage: Reduction by melatonin. Biochem Pharmacol [Internet] Elsevier , 1996 Apr 26; [cited 2017 Dec 1];51(8): 1095-9. Available from: https://www.sciencedirect.com/science/article/pii/000629529600055X

[26] Warfarin [Internet]. [cited 2017 Dec 1]. Available from: http://pmep.cce.cornell.edu/profiles/extoxnet/pyrethrins-ziram/warfarin-ext.html

[27] Mechanisms of Blood Coagulation [Internet]. [cited 2017 Dec 1]. Available from: http://departments.weber.edu/chpweb/hemophilia/mechanisms_of_blood_coagulation.htm

[28] Brodifacoum IPCS Inchem Data sheet on Pesticides No. 57 [Internet]. [cited 2017 Dec 1]. Available from: https://web.archive.org/web/20131213084637/http://www.inchem.org/documents/pds/pds/pest57_e.htm

[29] Kim S-Y, Kang J-Y, Hartman JH, Park S-H, Jones DR, Yun C-H, *et al.* Metabolism of R- and S-warfarin by CYP2C19 into four hydroxywarfarins. Drug Metab Lett [Internet] NIH Public Access , 2012 Sep 1; [cited 2017 Dec 1];6(3): 157-64. Available from: http://www.ncbi.nlm.nih.gov/pubmed/23331088

[30] Higdon J. Higdon J. Vitamin K, Linus Pauling Institute [Internet]. [cited 2017 Dec 1]. Available from: http://lpi.oregonstate.edu/mic/vitamins/vitamin-K

[31] Wilson D. Wilson D. Mechanisms of acute toxicity [Internet]. [cited 2017 Dec 1]. Available from: https://ntp.niehs.nih.gov/iccvam/meetings/at-wksp-2015/session4/1-wilson-508.pdf

[32]   Alcohol Overdose: The Dangers of Drinking Too Much [Internet]. [cited 2017 Dec 1]. Available from: https://pubs.niaaa.nih.gov/publications/AlcoholOverdoseFactsheet/Overdosefact.htm

[33]   Compound Interest - Lethal Doses of Water, Caffeine and Alcohol [Internet]. [cited 2017 Nov 29]. Available from: http://www.compoundchem.com/2014/07/27/lethaldoses/

[34]   Alcohol Poisoning [Internet]. [cited 2017 Dec 1]. Available from: http://www.intoxikon.com/Pubs/Facts on ALCOHOL POISONING_4_7_05.pdf

[35]   Material Safety Data Sheet, Ethyl Alcohol 200 Proof [Internet]. [cited 2017 Dec 1]. Available from: http://www.sciencelab.com/msds.php?msdsId=9923955

[36]   Levine M. Alcohol Toxicity [Internet]. [cited 2017 Dec 2]. Available from: https://emedicine.medscape.com/article/812411-overview

[37]   Alcohol Poisoning Deaths | VitalSigns | CDC [Internet]. [cited 2017 Dec 2]. Available from:https://www. cdc.gov/vitalsigns/alcohol-poisoning-deaths/index.html

[38]   Methanol Material Safety Data Sheet, Airgas [Internet]. [cited 2017 Dec 2]. Available from: https://wcam.engr.wisc.edu/Public/Safety/MSDS/Methanol .pdf

[39]   Ethylene Glycol Material Safety Dat Sheet. [Internet]. [cited 2017 Dec 2]. Available from: http://www.sciencelab.com/msds.php?msdsId=9927167

[40]   CDC - Carbon Monoxide Poisoning - Frequently Asked Questions [Internet]. [cited 2017 Dec 2]. Available from: https://www.cdc.gov/co/faqs.htm

[41]   Olas B. Carbon monoxide is not always a poison gas for human organism: Physiological and pharmacological features of CO. Chem Biol Interact [Internet] Elsevier , 2014 Oct 5; [cited 2017 Dec 2];222: 37-43. Available from: https://www.sciencedirect.com/science/article/pii/S0009279714002373

[42]   Material Safety Data Sheet MathesonTri Gas, Carbon Monoxide [Internet]. [cited 2017 Dec 2]. Available from: https://www.mathesongas.com/pdfs/msds/MAT04290.pdf

[43]   Brown OR. Oxygen, the Breath of Life [Internet] Bentham Science Publishers , 2017 [cited 2017 Dec 2]; Available from: https://ebooks.benthamscience.com/book/9781681084251/

* 9 7 8 1 6 8 1 0 8 6 9 8 9 *